P9-DHK-062

THE TALE OF THE DUELING NEUROSURGEONS

ALSO BY SAM KEAN

The Violinist's Thumb

The Disappearing Spoon

THE TALE OF THE DUELING NEUROSURGEONS

THE HISTORY OF THE HUMAN BRAIN
AS REVEALED BY TRUE STORIES
OF TRAUMA, MADNESS, AND RECOVERY

SAM KEAN

LITTLE, BROWN AND COMPANY
New York Boston London

Little, Brown and Company
Hachette Book Group
237 Park Avenue, New York, NY 10017
littlebrown.com

First Edition: May 2014

Little, Brown and Company is a division of Hachette Book Group, Inc.
The Little, Brown name and logo are trademarks of Hachette Book Group, Inc.

Illustrations by Andrew J. Brozyna, ajbdesign.com

Library of Congress Cataloging-in-Publication Data

Kean, Sam.
The tale of the dueling neurosurgeons : the history of the human brain as revealed by true stories of trauma, madness, and recovery / Sam Kean.—First edition.
p. cm.
Includes bibliographical references and index.
ISBN 978-0-316-18234-8 (hardcover)/ISBN 978-0-316-28648-0 (international)
I. Title.
1. Brain—physiology—Popular Works. 2. Neurosciences—history—Popular Works. 3. Brain Diseases—Popular Works. 4. Neurologic Manifestations—Popular Works. 5. Physicians—history—Popular Works.
RD593
617.4'80232—dc23 2014004910

10 9 8 7 6 5 4 3 2 1

RRD-C

Printed in the United States of America

The times have been
That, when the brains were out, the man would die
And there an end; but now they rise again.

WILLIAM SHAKESPEARE, *Macbeth*

Contents

Part V:
Consciousness

THE TALE OF THE DUELING NEUROSURGEONS

Rebus – n., a puzzle that involves piecing together pictures, letters, and sounds to form a hidden word or phrase. For example,

B + = Brain

N.B.: I've included a rebus at the start of each chapter, to highlight that chapter's theme and content. If you succeed in decoding all thirteen, drop me a message at http://samkean.com/contact-me, and brag a little. :) Or, if you get stuck, go ahead and e-mail me anyway, and I can give you a hint…

PART I

GROSS ANATOMY

Introduction

I can't fall asleep on my back—or rather, I don't dare to. In that position I often slip into a fugue state where my mind wakes up from a dream, but my body remains immobile. In this limbo I can still sense things around me: sunlight trickling through the curtains, passersby on the street below, the blanket tented on my upturned feet. But when I tell my body to yawn and stretch and get on with the day, nothing happens. I'll recite the command again—*Move, you*—and the message echoes back, unheeded. I fight, I struggle, I strain to twiddle a toe or flex a nostril, and it does no good. It's what being reincarnated as a statue would feel like. It's the opposite of sleepwalking—it's sleep paralysis.

The worst part is the panic. Being awake, my mind expects my lungs to take full, hearty breaths—to feel my throat expanding and my sternum rising a good six inches. But my body—still asleep, physiologically—takes mere sips of air. I feel I'm suffocating, bit by bit, and panic begins to smolder in my chest. Even now, just writing this, I can feel my throat constrict.

As bad as that sounds, some sleep paralytics have it worse. My episodes don't last that long: by concentrating all my energy, Zen-master-like, on twitching my right pinky, I can usually break the trance within a few minutes. Some people's episodes drag on for hours, full nights of torture: one Korean War vet reported feeling more terror during a single episode of sleep paralysis than during his entire thirteen months of combat. Other people nod off narcoleptically and slip into this state during the day. One poor woman in England has been declared dead three times and once woke up in a morgue. Still other people have out-of-body experiences and feel their spirits careening around the room. The unluckiest ones perceive an evil "presence"—a witch, demon, or incubus—pressing down on their necks, smothering them. (The very "mare" in nightmare refers to a witch who delights in squatting on people's chests.) Nowadays people sometimes weave this feeling of paralysis into alien abduction stories; presumably they're strapped down for probing.

Sleep paralysis doesn't actually open a portal into the supernatural, of course. And despite what I may have thought when young, sleep paralysis doesn't offer proof of dualism, either: the mind cannot appear outside the body, independent of it. To the contrary, sleep paralysis is a natural by-product of how our brains work. In particular, it's the by-product of faulty communication among the three major parts of the human brain.

The base of the brain, including the brainstem, controls breathing, heart rate, sleeping patterns, and other basic bodily functions; the brainstem also works closely with the nearby cerebellum, a wrinkly bulb on the brain's derriere that helps coordinate movement. Together, the brainstem and cerebellum are sometimes called the reptile brain, since they function approximately like the brain of your average iguana.

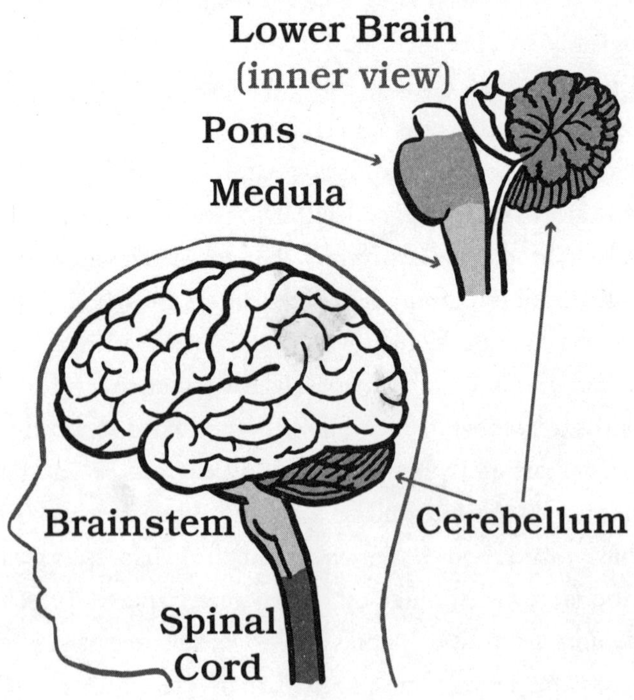

The second part, the so-called mammal brain, sits deep inside the skull, just north of the brainstem. The mammal brain relays sensory input around; it also contains the limbic system, which helps capture memories, regulate emotion, and distinguish between pleasant and rotten experiences. Unlike the instinct-driven reptile brain, the mammal brain can learn new things quite easily. To be sure, some neuroscientists deride the mammal/reptile division as too simplistic, but it's still a useful way to think about the brain's lower regions.

Middle Brain/Limbic System

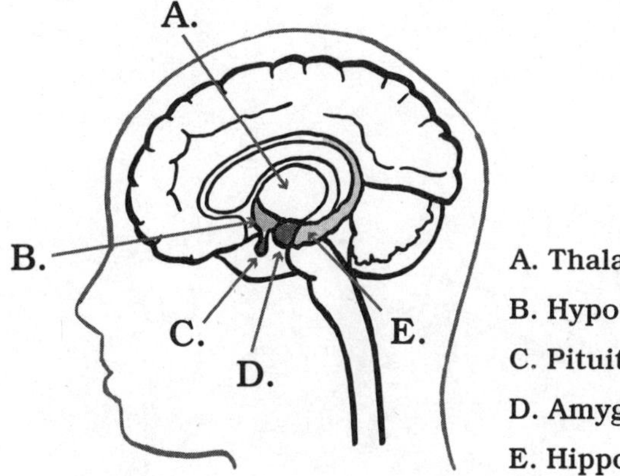

A. Thalamus
B. Hypothalamus
C. Pituitary gland
D. Amygdala
E. Hippocampus

Both of these nether regions control automatic processes, things we don't think about, or want to. This autopilot frees up the outermost part of the brain, the primate brain, for advanced duties, especially in humans. We can further divide the wrinkly primate brain into four lobes: the frontal lobes (near the front of the brain), which initiate movement and help us plan, make decisions, and set goals; the occipital lobes (back of the brain), which process vision; the parietal lobes (atop the brain, the pate), which combine vision, hearing, touch, and other sensations into a "multimedia" worldview; and the temporal lobes (side of the brain, behind the temples), which help produce language, recognize objects, and link sensations with emotions.

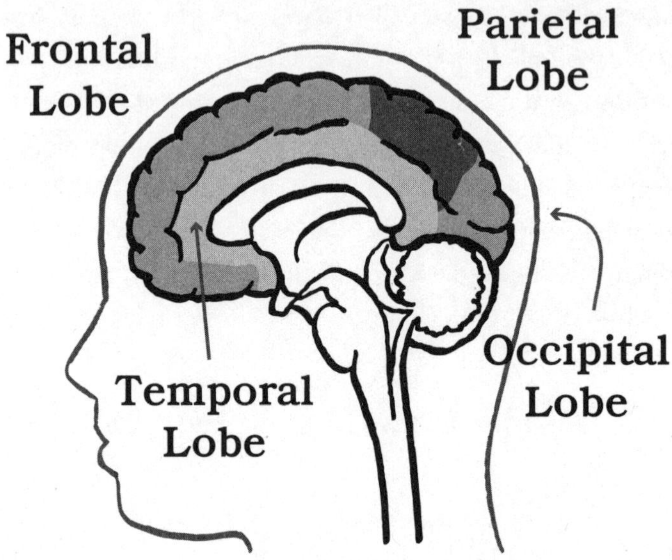

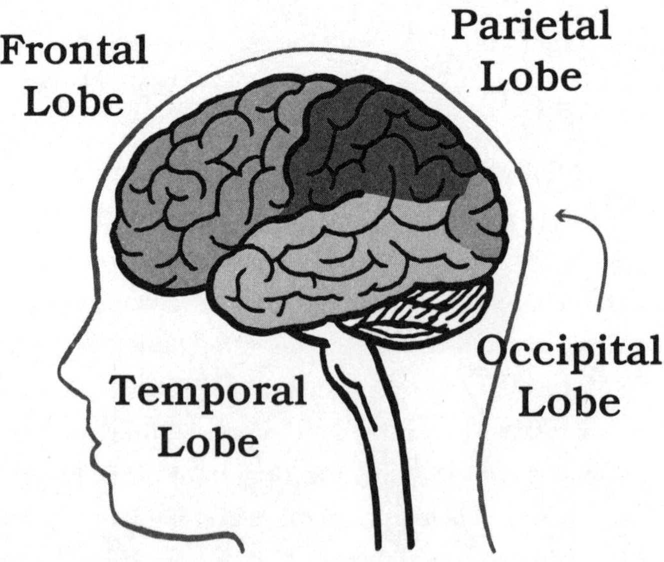

The reptile, mammal, and primate brains constantly exchange messages, usually via chemicals, and their various internal structures work together almost seamlessly. Almost.

Deep inside the reptile brain sits the pons, a hump in the brainstem an inch long. When we fall asleep, the pons initiates dreaming by sending signals through the mammal brain to the primate brain, where dreams stir to life. During dreams, the pons also dispatches a message to the spinal cord beneath it, which produces chemicals to make your muscles flaccid. This temporary paralysis prevents you from acting out nightmares by fleeing the bedroom or taking swings at werewolves.

While mostly protective, this immobility sometimes backfires. Sleeping on your back can collapse the airways in your throat and deprive the lungs of oxygen. This isn't a huge deal during nonparalyzed, nondream sleep: the parts of the brain that monitor oxygen levels will rouse your body a little, halfway to waking, and you'll snort, shift your head, or roll over. To get oxygen during dream sleep, though, the brain has to order the pons to stop paralyzing your muscles. And for whatever reason—a chemical imbalance, a frayed neural wire—the pons doesn't always obey. So while the brain succeeds in rousing the mind a little, it can't turn off the spigot for the paralysis chemicals, and the muscles remain limp.

Things go south from there. If this limbo persists, the mind wakes up fully and, sensing something amiss, trips a circuit that includes the amygdala, a structure in the mammal brain that amplifies fear. A fight-or-flight response wells up—which exacerbates the problem, since you can't do either. This is when the panic starts. And again, some people have it much worse. At least with me, the actual dream I'm having stops as soon as my mind wakes up. Not so in some people: they never quite escape the dream state. They're semialert to their surroundings, they're paralyzed, *and* their brains keep conjuring up dream nonsense. Because the human mind is quite good at making spurious connections, they then link the characters in these hallucinations to their paralysis, as if one caused the other. It's no wonder some people believe in demons and aliens: they actually see and feel them.

So, yeah, there's a reason I don't sleep on my back anymore. But even though I dreaded the experience, sleep paralysis did teach me something valuable about the brain: that everything is interconnected. Starting with nothing but chemicals way down deep in the reptile parts, I could nevertheless—if I followed the tumbling dominoes far enough and patiently worked my way up from chemicals to cells to circuits to lobes—gain insight into the most rarified realm of the human mind, a belief in the supernatural. One little brain malfunction could be parlayed into so much more.

In fact, the more I read about neuroscience and the interplay of different neural structures, the more I realized that this huge yield wasn't unusual. Tiny flaws in the brain had strange but telling consequences all the time. Sometimes these flaws wipe out general systems like language or memory. Other times, something very specific dies. Destroy one small node of neurons, and people lose the ability to recognize fruits and vegetables but not other food. Destroy another node and they lose the ability to read—even though they can still write. Still other malfunctions tack a phantom third arm onto someone's torso, or convince her that the very hand on the end of her arm belongs to someone else. Overall, these flaws reveal how the brain evolved and how it's put together, and I realized that you could write a whole natural history of the brain from just such cases...

~

Until the past few decades, neuroscientists had one way to plumb the human brain: wait for disaster to strike people and, if the victims pulled through, see how their minds worked differently afterward. These poor men and women endured strokes, seizures, saber gashes, botched surgeries, and accidents so horrific—like having a four-foot iron javelin driven through the skull—that their survivals seemed little short of miracles. To say these people "survived," though, doesn't quite capture the truth. Their bodies survived, but their minds didn't quite; their minds were warped into something new. Some people lost all fear of death; others started lying incessantly; a few became pedophiles. But however startling, in one way these transformations proved predictable, since people with the same deficit tended to have damage in the same area of the brain — offering vital clues about what those areas did. There are a thousand and one such stories in neuroscience, and *The Tale of the Dueling Neurosurgeons* recounts the best of them, resurrecting the lives of the kings, cannibals, dwarfs, and explorers whose struggles made modern neuroscience possible.

Many of these people's lives are inherently dramatic, because their ailments felled them within days, even minutes. And as far as possible, rather than just recite the details of doctors' visits or provide a litany of one damn brain-scan study after another, this book enters into the minds of victims, to give you a sense of what it's like actually *living* with crippling amnesia or the conviction that all your loved ones have been replaced by imposters. While some of the stories have familiar characters (it's probably illegal to write about neuroscience nowadays without mentioning H.M. or Phineas Gage), many characters will be new. Even with some of the standbys, like Gage, much of what you "know" is probably wrong. Not all the stories are tragic, either. Some are plain enchanting, like those about people whose senses fuse together in trippy ways, so that odors make noises and textures produce flashes of color. Some are uplifting, like tales of blind people who learn to "see" their surroundings through batlike echoes. Even the stories about accidents are, in many cases, stories of triumph, stories about the brain's resiliency and ability to rewire itself. And these tales remain relevant to neuroscience today: despite the (often overhyped) advances of fMRI and other brain-scanning technologies, injuries remain the best way to infer certain things about the brain.

In general, each chapter here recounts one narrative tale; that's how the human brain remembers information best, in story form. But beneath these ripping yarns there are deeper threads, threads that run through all the chapters and bind them together. One thread concerns scale. The early chapters explore small physical structures like cells; think of these sections like individual red and green and yellow fibers to feed into a loom. With each successive chapter we'll cover larger and larger territories, until we can see the full Persian carpet of the brain. Another thread concerns neural complexity. Every chapter adds a little more ornament to the rug, and the motifs and themes of the early chapters get repeated later on, allowing you to see the brain's intricate, interlocking patterns more clearly the closer you look, with each passing page.

The book's first section, "Gross Anatomy," familiarizes you with the brain and skull, providing a map for future sections. It also shows the genesis of modern neuroscience from one of the most important cases in medical history.

"Cells, Senses, Circuits" delves into the microscopic phenomena that ultimately underlie our thoughts, things like neurotransmitters and electrical pulses.

"Body and Brain" builds upon those smaller structures to show how the brain controls the body and directs its movement. This section also shows how bodily signals like emotions bend back and influence the brain.

"Beliefs and Delusions" bridges the physical and mental, showing how certain defects can (à la sleep paralysis) give rise to tenacious, and pernicious, delusions.

Finally, all these sections build toward the last section, "Consciousness," which explores memory and language and other higher powers. This includes our sense of self—the "inner you" we all carry around in our heads.

By book's end you'll have a good sense of how all the different parts of your brain work, and especially how they work together. Indeed, the most important theme in this book is that you can't study any part of the brain in isolation, no more than you can hack the Bayeux Tapestry up and still grasp its intricacies. You'll also be prepared to think critically about other neuroscience you read about and to understand future advances.

Above all, I wrote *The Tale of the Dueling Neurosurgeons* to answer a question, a question that has clawed at me ever since those first scary episodes of sleep paralysis: where does the brain stop and the mind start? Scientists have by no means answered this question. How a conscious mind emerges from a physical brain is still the central paradox of neuroscience. But we have some amazing leads now, thanks largely to those unwitting pioneers—those people who, usually through no fault of their own, suffered freak accidents or illnesses and essentially sacrificed a normal life for the greater good. In many cases what drew me to these stories was the very commonness of their heroes, the fact that these breakthroughs sprang not from the singular brain of a Broca or Darwin or Newton, but from the brains of everyday people—people like you, like me, like the thousands of strangers we pass on the street each week. Their stories expand our notions of what the brain is capable of, and show that when one part of the mind shuts down, something new and unpredictable and sometimes even beautiful roars to life.

CHAPTER ONE

The Dueling Neurosurgeons

One of the landmark cases in medical history involved King Henri II of France, whose suffering foreshadowed almost every important theme in the next four centuries of neuroscience. His case also provides a convenient introduction to the brain's layout and general makeup.

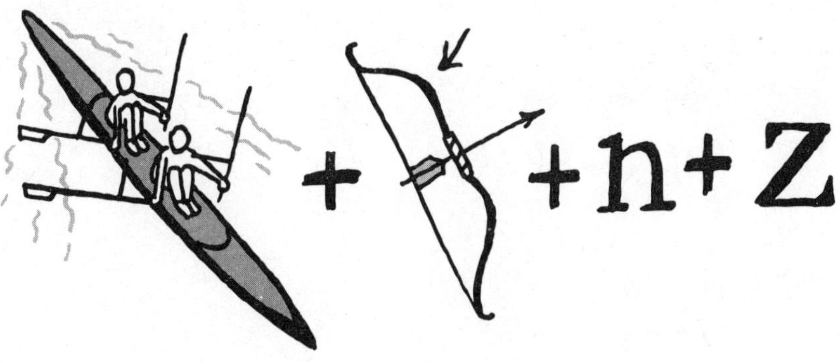

The world would have looked stunningly, alarmingly bright to the king of France, then suddenly dark. During the charge, little light penetrated the cocoon of his helmet. Darkness was safety. But when the visor was wrenched open, the sunlight punched his eyes, a slap as sharp as a hostage would feel the moment the bag was torn off his head. In his last split second of normal life, Henri's eyes might have registered a glimpse of the scene in front of him—the glint of sand kicked up by his horse's hooves; the throbbing white ribbons wrapped around his lance; the glare off the armor of his charging opponent. As soon as he was clobbered, everything dimmed. Just a handful of doctors in the world in 1559 could have foreseen the damage already diffusing through his skull. But even these men had never worked on a case so important. And over the next eleven days, until King Henri was past danger, most of the great themes of the next four centuries of neuroscience would play themselves out in the microcosm of his brain.

The unlikely king, unlikely queen, and unlikely royal mistress were celebrating a supposed end to violence that day. Queen Catherine looked like royalty itself in a gown of silk interwoven with gold fibers, but she'd actually grown up an orphan. As a fourteen-year-old in 1533, she'd watched helplessly as her family, the Medici of Florence, negotiated her marriage to an unpromising prince of France. She'd then endured a decade of barrenness with Henri before saving her life by squeezing out two heirs. And throughout it all, she'd had to endure the rivalry of her cousin Diane. Diane de Poitiers had been married to a man forty years her senior until just before Catherine's arrival in Paris. When he died, Diane donned black and white (French mourning colors) in perpetuity, a show of piety. Yet this thirty-five-year-old beauty lost no time in turning cougar on the fifteen-year-old Prince Henri, first enslaving him with sex, then parlaying this hold over him into real political power, much to the queen's disgust.

Le Roi, Henri II, had never been groomed for the throne; he'd become heir apparent only when his handsomer and more charming older brother died after a game of tennis. Henri furthermore had a tough early reign. Paranoid about Protestant spies, he'd started chopping off the tongues of "Lutheran scum" and burning them at the stake, making himself hated throughout France. He'd also prolonged a series of stupefyingly complex wars with Spain over Italian territories, bankrupting the realm. By the late 1550s Henri owed forty-three million livres to creditors—over twice his yearly income—with some loans at 16 percent interest.

So in 1559 Henri abruptly brought peace to France. He signed a treaty with Spain, and although many (including Catherine) fumed over Henri's giving away Italy, he stopped the ruinous military campaigns. Two important clauses in the treaty established alliances through marriages—an immediate wedding for Henri and Catherine's fourteen-year-old daughter to the king of Spain, and a second for Henri's spinster sister to an Italian duke. To celebrate the marriages, Henri organized a five-day jousting tournament. He had to borrow two million livres more, but workmen spent May and June ripping up cobblestones and packing down sand near Henri's palace in Paris to make a jousting list. (Protestants awaiting punishment in nearby dungeons could hear the clamor in their cells.) A few weeks before the tournament, carpenters erected some rickety timber galleries for royal guests and draped them with standards and banners. On the day of, peasants climbed onto rooftops to point and holler.

On the third day of festivities, a Friday, June 30, Henri himself decided to joust. Despite the heat he wore fifty pounds of gold-plated armor adorned with Diane's colors, mostly black-and-white swirls. Whatever his faults, Henri looked regal upon a horse, and he entered the list on a handsome, chestnut-colored steed. During his first run he unmanned (which is to say, unhorsed) his future brother-in-law with a blow from his lance; a short while later he unmanned a local duke, knocking him onto his arse as well. When young, Henri had had a reputation as a brooding sort, but he was in high spirits that day, and arranged for a third and final joust against a powerful young Scotsman, Gabriel Montgomery.

The king and Montgomery put perhaps a hundred yards between them, and when a trumpet sounded, they took off. They clashed—and Henri got his bell rung. Montgomery bludgeoned him just below the neck, and Henri lost a stirrup and nearly careened off his horse.

Embarrassed, the king wheeled around and announced that "we" would tilt with Montgomery again—a bad idea for any number of reasons. It violated the laws of chivalry, as he'd already jousted the maximum three times. It also spooked his court. Catherine had dreamed the night before of Henri lying facedown in blood, and two of her astrologers had already prophesied the king's doom. (One of them, Nostradamus, had written a quatrain four years earlier that read, "The young lion overcomes the old / on the field of war in single combat. / He pierces his eyes in a cage of gold. / Two wounds one, then dies a cruel death.") Unnerved, Catherine sent a messenger to warn Henri off.

Finally, Henri had been suffering from vertigo and headaches recently, and his attendants found him shaken after his latest joust. Cruelly, though, a blow to the head can cloud someone's judgment when he needs it most, and like a linebacker or boxer of today, Henri insisted on jousting again. Montgomery demurred, and the crowd watched in embarrassment as Henri berated him and challenged him—on his allegiance, before God—to tilt again. At 5 p.m. they lined up. Some eyewitnesses later claimed that an attendant fastened the king's visor improperly. Others said that Henri wiped his brow and, in his fog, forgot to refasten the visor. Still others insisted that he cocked it up, in spite. Regardless, this time Henri didn't wait for the trumpet before charging.

During a joust, a low timber fence separated the combatants, and they charged each other left shoulder to left shoulder, shield hand to shield hand. They held their fourteen-foot wooden lances in their right arms and had to angle them across their bodies to strike. A proper blow therefore not only jolted but twisted the opponent, and the force often broke the lance. Sure enough, the king's lance shattered when it met Montgomery, and Montgomery's lance exploded into splinters when it struck the king just below the neck. Both men jerked, and the courtiers in hose and doublets, the women adorned with ostrich feathers, the peasants hanging on the eaves, all of them whooped at the teeth-rattling blow.

The action, though, was not over. Given the commotion, no one quite knows what happened next. Perhaps Montgomery's broken shaft buckled upward like an uppercut, or perhaps a splinter of wood leapt up like shrapnel. But somewhere in the melee, something knocked open the king's gold-plated visor.

Now, many contemporaries blamed Montgomery for what happened next, because the moment his lance splintered he should have flung it aside. But the brain can react only so quickly to stimulus—a few tenths of a second at best—and a brain fogged from jousting would have responded more slowly still. Besides, Montgomery had an awful momentum, and even as the crowd's roar lingered, his horse took another gallop. An instant later the jagged lance butt in his hand struck the king dead between his eyebrows. It raked across his naked face, wrenching his skull sideways and digging into his right eye. *He pierces his eyes in a cage of gold.*

But Nostradamus spoke of two wounds, and a second, deeper wound, to Henri's brain, proved worse. Compared to those of most mammals, the four lobes of the human brain are grotesquely swollen. And while our skulls provide some good protection, the very hardness of the cranial bones also poses a threat, especially since the skull is surprisingly jagged on the inside, full of edges and ridges. What's more, the brain actually floats semifreely inside the skull; it's attached to the body really only at the bottom, near the stalk of the brainstem. We do have cerebrospinal fluid between the skull and brain to buoy and cushion it, but the fluid can absorb only so much energy. During impact, then, the brain can actually slide counter to the skull's motion and slam into its bones at high speed.

As the butt of Montgomery's lance struck home, Henri would have felt both a blow and a twist, like a mean hook to the jaw. The blow likely sent a small shock wave through his brain, a ripple of trauma. The rotational force was likely even worse, since torque stresses the brain unequally at different points, tearing at its soft seams and opening up thousands of microhemorrhages. Henri, an expert equestrian, nevertheless kept his saddle after the impact: the muscle-memory circuits in his brain kept him balanced and kept his thighs squeezing the horse. But on a deeper level, the twist and the blow tore open millions of neurons, allowing neurotransmitters to leak out and flood the brain. This would have caused untold numbers of other neurons to fire in panic, a surge of electrical activity reminiscent of a mini-seizure. Although few men of science believed in such things, at least one doctor in Paris knew that Henri had suffered a mammoth concussion.

After the clash Montgomery yanked his horse's reins and whirled to see what he'd done. Henri had slumped down onto the neck of his Turkish steed, a horse forevermore known as Malheureux, *unlucky.* But however unlucky, Malheureux was disciplined, and when it felt its reins slacken upon Henri's collapse, it kept galloping. The now-unconscious king bobbed on his horse's back as if keeping time, his visor clanging down on the shards of wood protruding from his eye.

~

The two greatest doctors in Europe would soon converge on the king, but before they could do so, courtiers and sycophants of all stripes poured out of the stands toward Henri, each one craning for a glimpse and calculating whether his fortunes would rise or fall if Le Roi died. To most observers the entire French monarchy now looked as rickety as the timber grandstands. The dauphin (the heir apparent) was a frail, milquetoast boy of fifteen; he fainted at the mere sight of Henri's injury. The shaky truce between Catherine and Diane depended entirely on Henri's living, as did the false peace between other political factions. The two royal weddings, not to mention the peace of Europe, threatened to unravel as well.

Eased down from his horse, Henri lay stunned. Montgomery pushed to the front of the crowd to beg, somewhat incongruously, that the king both forgive him and also cut off his head and hands. Upon surfacing to consciousness, the king instead absolved him, without beheading or behanding. Henri drifted in and out after that, and finally insisted on rising and walking (albeit with support) up the palace steps to his bedroom. His physicians set about removing a four-inch splinter from his eye, but had to leave many smaller ones in place.

Among those doctors attending the king was Ambroise Paré. A thin, prim man, Paré served as royal surgeon—a job less prestigious than it sounds. The son of a cabinetmaker, Paré hailed from a village

Dueling neurosurgeon Ambroise Paré.
(National Library of Medicine)

in north France, where he'd trained as a "barber-surgeon." In short, barber-surgeons cut things, which differentiated them from proper physicians. He might start his day at 6 a.m. shaving beards and trimming periwigs, then amputate a gangrenous leg after lunch. In the early 1200s, the Catholic church had declared that no proper Christians, including physicians, could shed blood; physicians therefore looked down upon surgeons as butchers. Early in his career Paré had stood below even most surgeons because he spoke no Latin. Nor could he afford his licensure fee, so he became a battlefield surgeon at twenty-six, joining the army as a ragtag follower with no rank or regular salary. Injured soldiers paid what they could, be it casks of wine, horses, a half crown, or (sometimes) diamonds.

Paré took to soldiering like a house afire, hobnobbing with generals by day and getting drunk with lower-ranked officers at night. Over the next thirty years he worked in seventeen campaigns across Europe. But he made his first important discovery while a tyro. Most doctors in the early 1500s considered gunpowder poisonous, and cauterized any bullet wound, however slight, by dousing it with boiling elderberry oil. To his mortification, Paré ran out of elderberry oil one night after a battle. Begging their forgiveness, he patched up his patients' wounds with a paste of egg yolk, rose water, and turpentine instead. He expected every one of these "untreated" soldiers to die, but they were fine the next morning. In fact they were thriving compared to those treated with boiling oil, who writhed in pain. Paré realized he'd effectively run an experiment, with astounding results, since his trial group had fared much better than his controls.

That morning changed Paré's entire outlook on medicine. He refused to use boiling oil ever again, and set about perfecting his egg/turpentine paste instead. (Over the years the recipe changed somewhat, and eventually included earthworms and dead puppies.) On a deeper level, that morning taught Paré to experiment and to observe results for himself, no matter what ancient authorities said. It was a symbolic conversion, really: by abandoning boiling oil—with all its medieval overtones—Paré effectively abandoned a medieval mentality that accepted medical advice on faith.

As is evident from his case reports, Paré lived in an era of near-cartoonish violence: one day might find him treating a twelve-year-old girl mauled by the king's pet lion; the next might find him literally standing on a duke's face, to get enough leverage to yank a broken spearhead out. But Paré handled it all with aplomb, and his willingness to experiment made him an innovative surgeon. He developed a new drill-and-saw contraption to "trepan" the skull—that is, open a hole in the bones and relieve pressure on the brain, whether from inflammation or fluid buildup. He also developed tests to distinguish—in particularly gory head wounds—between fat, which was harmless to scrape away, and oozing bits of fatty brain tissue, which weren't. (In short, fat floats on water, brain sinks; fat liquefies in a frying pan, brain shrivels.) When describing a patient's recovery, Paré usually pooh-poohed his own role: "I treated him, God healed him," he famously said. But his many near resurrections earned Paré quite a reputation, and Henri eventually appointed him "surgeon to the king."

Despite his expertise with head wounds, Paré still stood below the king's physicians in the medical hierarchy, and he deferred to them in the busy hours after the jousting disaster. The physicians force-fed Henri a potion of rhubarb and charred Egyptian mummy (a treatment Paré rolled his eyes at in private), and they opened the king's veins to bleed him, even as he bled spontaneously from his colon. The English ambassador noted the king had a "very evil rest" that first night, but most of the attending doctors remained optimistic that he'd suffered little damage beyond his right eye. And in fact, when the king surfaced into consciousness the next morning, he seemed to have his wits about him.

But Henri soon had to face the fact that Catherine had effectively seized control of France. He asked after Montgomery, and frowned to learn that the Scotsman, not trusting Catherine, had already fled.

Henri called out for his mistress, but Catherine had stationed soldiers at the palace door to bar Diane's entrance. Perhaps most startlingly, Henri learned that Catherine had ordered the decapitation of four criminals—then bade Henri's doctors to experiment on their heads with Montgomery's broken lance stump, to devise a strategy for treatment.

Meanwhile, a messenger on horseback was sprinting northeast through forest and field to Brussels, bound for the court of Philip II, the king of Spain. (Confusingly, Spanish kings lived in northern Europe, in conquered territory.) Although the recent peace treaty had secured Philip's marriage to Henri's daughter, Philip hadn't deigned to attend his own wedding in Paris, explaining that "kings of Spain do not run after brides." (Philip sent a proxy, a duke, to stand in at the ceremony instead. To legally "consummate" the marriage, the duke approached the princess's bedchamber that night, removed his boot and hose, and slipped his foot beneath the bedcovers to caress the girl's naked thigh. There was much ribald speculation in Paris about whether this satisfied her.*) Despite his haughtiness toward Henri's family, Philip wanted Henri to live, and soon after the messenger arrived, Philip roused his best physician, the one man in Europe whose expertise with the brain rivaled Paré's.

As a teenager in Flanders, Andreas Vesalius had dissected moles, mice, cats, dogs, and whatever other animals he could snare. But animal dismemberment couldn't quite sate him, and before long he began pursuing his real passion, human dissection. He began robbing graves at midnight, sometimes fighting wild dogs over scraps. He also let himself get locked outside the city walls at night to rob skeletons from the gallows, clambering up thirty-foot gibbets to cut down swaying pickpockets and murderers, counting himself lucky if crows hadn't refashioned their anatomies too much already. He smuggled

---

*This and all upcoming asterisks refer to the Notes and Miscellanea section, which begins on page 359 and goes into more detail on various interesting points.

Dueling neurosurgeon Andreas Vesalius.
(National Library of Medicine)

the cadavers back into the city beneath his garments, then stored them in his bedroom for weeks, to linger over their dissections like a cannibal gourmand over a meal. He delighted in clutching every organ, too, even crushing them between his fingers to see what oozed out. However creepy, his obsession revolutionized science.

Vesalius eventually enrolled in medical school, and like everyone else in the previous thirteen centuries, his medical training basically consisted of memorizing the works of Galen, a physician born in AD 129. Human dissection was taboo back then, but luckily for him, Galen served as a doctor for Roman gladiators, which was just about the best training possible for an anatomist: gladiator wounds could get pretty gnarly, and he probably saw more human innards than anyone alive then. He soon founded a school of anatomy, and his work was so innovative and all-encompassing that he stunted the field, since his small-minded followers couldn't evolve past him. By the Renaissance, the birth pangs of a new science of anatomy had begun, but most "anatomists" still cut into the body as little as possible. Anatomy lectures were similarly a joke: they mostly consisted of an expert sitting on a throne and reciting Galen aloud while, beneath him, a lowly barber hacked open animals and held their greasy entrails up. Anatomy was theory, not practice.

Vesalius—a swarthy man with a virile black beard—adored Galen, but after immersing himself in human flesh, he began to notice discrepancies between the gospel of Galen and the evidence on the dissecting table. At first Vesalius refused to believe his own eyes, and told himself that he must have cut into some anomalous bodies. He even entertained the theory that the human body had changed since Galen's time, possibly because men now wore tight trousers instead of togas. Eventually, though, Vesalius admitted he was grasping, and that, however unthinkable it seemed, Galen had erred. Around 1540 he compiled a list of two hundred howlers, and determined from them that Galen had supplemented his gladiatorial work by dissecting sheep, apes, oxen, and goats, and then extrapolated to humans. This bestiary left human beings with extra lobes on the liver, a two-chambered heart, and fleshy "horns" on the uterus, among other mutations. Galen's shortcomings became glaringly obvious when Vesalius probed the brain. Galen had dissected mostly cow brains, cow brains being large and abundant in the butcher stalls of Rome. Unfortunately for Galen, humans have vastly more complex brains than cows, and for thirteen hundred years, physicians were left trying to explain how the brain worked based on a faulty notion of how it was put together.

Vesalius vowed to reform the science of anatomy. He began calling out, even exposing, prominent "anatomists" who never bothered dissecting bodies themselves. (About one, Vesalius sneered that he'd never seen the man with knife in hand, except when carving mutton at dinner.) More important, Vesalius reached a wider audience by

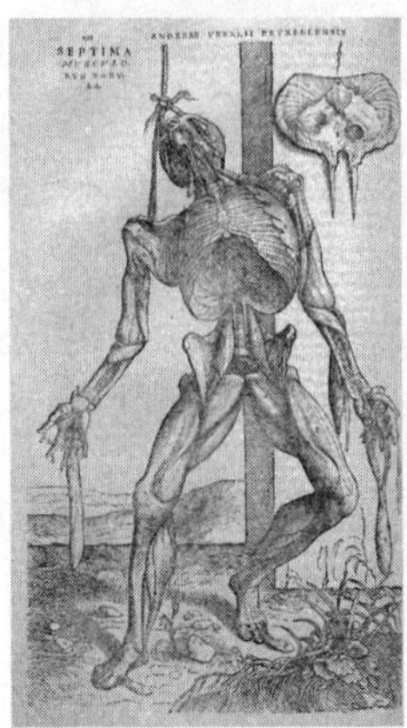
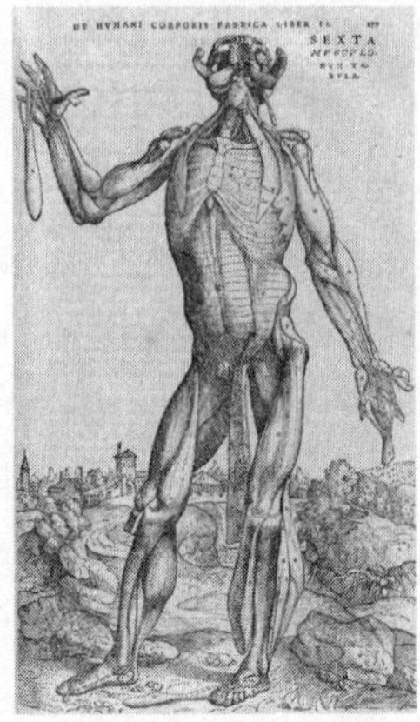
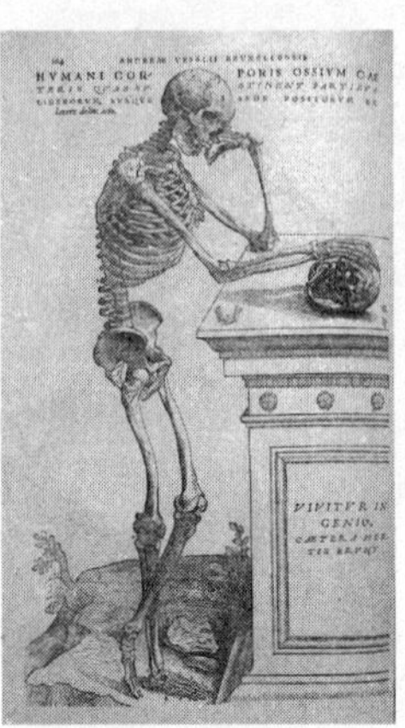

Drawings from Andreas Vesalius's *On the Fabric of the Human Body,* one of the most beautiful scientific books ever published. (National Library of Medicine)

composing one of the cherished works of Western civilization, *De Humani Corporis Fabrica* (On the Fabric of the Human Body).

Beyond the few crude diagrams in other books, this was the first anatomy text to include realistic drawings of the human form. And what drawings these were. Vesalius sought out the best local artist to illustrate his magnum opus, and since he was working in Padua then, this happened to be Titian, whose school of artists soon brought Vesalius's vision of the human form to life. Unlike in modern textbooks, the bodies in *Fabrica* don't lie flat and lifeless on a table. They rise and strut and pose like classical statues. Some do a veritable striptease with their flesh, peeling back layer after layer to reveal their inner organs and organic essence. In darker scenes, bodies sway from ropes or clasp their hands together in agonized prayer. One skeleton digs his own grave; another contemplates a skull, alas-poor-Yorick-like. Titian's apprentices labored over even the backgrounds of the pictures, planting the cavorting cadavers in the lovely, rolling landscapes near Padua. As in that era's painting and sculpture, the realism was unsurpassed, making *Fabrica* one of the greatest marriages of art and science* ever produced. Pictures in the book's seventh and crowning volume, on the brain and related structures, distinguished scores of important details for the first time. Other anatomists had passed their eyes over the brain, but in a literal sense Vesalius, like a great artist, was the first person to really see it.

Ever obsessive, Vesalius agonized over every detail of *Fabrica,* including what paper and font to use, and he crossed the Alps from Italy to Switzerland to oversee its printing. For the first bound copy, he found another artist to hand-paint the drawings and, after cladding the book in purple silk velvet, carried it farther north and presented it to the Holy Roman Emperor, Charles V. It was June 1543, and in a remarkable coincidence, Nicolaus Copernicus had published *On the Revolutions of the Heavenly Spheres* a week before. But whereas *On the Revolutions,* written by a seventy-year-old astronomer, demoted human beings from the center of the cosmos, *Fabrica,* written by a twenty-eight-year-old anatomist, elevated us, celebrating us as architectural marvels. This near-pagan glorification of the body didn't please everyone, not even some anatomists, who vilified Vesalius and demanded that he retract every criticism of Galen. (Vesalius's former mentor smeared him as *Vesanus,* Latin for "madman," with a pithy anatomical pun attached to its posterior.) Being ignorant of medical matters, Charles V adored *Fabrica* and promoted Vesalius to court physician.

By 1559, though, Charles had died, and Vesalius found himself serving in the court of Charles's son, the cold and remote Philip. He spent most days treating nobles for gout and VD and bowel obstructions, with little time for original work. So when word of Henri's disastrous joust arrived, Vesalius jumped, dashing down to Paris on a relay of mail carriages, covering two hundred miles in forty-eight hours.

He soon met Paré, and modern neuroscientists sometimes bite their knuckles at the thought of this encounter—two titans, meeting at last! They'd actually almost met before, in 1544, near Saint-Dizier, when the army Vesalius was serving with laid siege to Paré's army.

This time around, any "combat" would be mano a mano, and these two proud and ambitious men likely circled each other, sizing the other up. But they had little time to waste in posturing.

If contemporary sketches are accurate, the king's bedchamber had deteriorated into a zoo. Dogs ran about, apothecaries chopped up herbs and mummy bits at the foot of the bed, and courtiers circled like vultures, interrupting Henri's rest. Henri lay on a four-poster bed with sumptuous blankets and a nude bust perched above the headboard. Case notes report that his face had swelled grotesquely, and his neck had stiffened like old French bread. His left eye could still see, but the lance had blinded the right and exposed the bone around the socket; the pus-stained bandage there surely clashed with the silk pillows. Given modern knowledge of brain trauma, we can surmise that Henri had a metallic taste in his mouth. Worst of all he could no doubt feel some dim black thundercloud, a massive headache, throbbing in the back of his skull. In his lucid moments, Henri gamely conducted state business, dispatching letters, arranging for his sister's marriage to proceed, even condemning some Lutheran scum. But as his brain swelled and the headache spread, he grew confused, and his vision came and went. He slept fitfully, and asked repeatedly for soothing music, which was never denied him, and for Diane, who was.

Miraculously, Paré and Vesalius found no fractures on Henri's skull, not even a hairline crack. (Since ancient times, doctors had a few ways of searching for cracks. They might dab ink on the top of the head and watch whether it seeped through, or they might thwack the skull with a stick and listen, since cracked and intact skulls sound different, much like cracked and intact bells do.) Many court physicians rejoiced at this news and proclaimed that Henri would therefore live: like most doctors then, they believed that the brain could not suffer any serious damage in the absence of a skull fracture, much like an egg yolk can't be damaged without the shell being cracked. (Some jurisdictions didn't even recognize a blow to the head as murder unless it broke the skull.) And admittedly, skull fractures did *look* bloody awful, much more sickening than nonfractures, so the reasoning made some sense.

Vesalius and Paré reasoned differently. Upon meeting the king, Vesalius produced a white cloth and asked Henri to bite down on it. Rather irreverently, he then ripped it from the royal mandible. Henri's body convulsed, his hands shot to his head, he howled in pain. You can imagine the sound of a dozen swords being unsheathed at this affront, but the stunt convinced Vesalius that Henri would die. The author of *Fabrica* knew better than anyone how delicate the brain is—you can scoop it with a spoon, like ripe avocado—and long experience told him that people with pain so intense usually didn't survive.

For his part Paré drew on battlefield experience. Not infrequently a soldier beaned by a shell or cannonball would betray no external symptoms—he might not even bleed. But his mind would wax and wane, and his brain would soon shut down. To probe this mystery, Paré would perform a quick autopsy. Autopsies were rare and usually illegal back then, but such laws were relaxed on the battlefield. And when Paré did his furtive autopsies, he often found swollen and bruised and sometimes even dead tissue inside these brains—signs of a controversial new diagnosis called a concussion. Paré had also seen cases where the head took a blow on one side but the brain damage was concentrated on the opposite side—a so-called contrecoup injury. These were in fact often the deadliest injuries. So in a prediction to outdo even Nostradamus, Paré suggested that Henri's brain had suffered a mortal contrecoup concussion, with damage localized in the back. Each man drew on different expertise in judging the king a goner, but they both disregarded the ancient imperative about gory head injuries necessarily being the worst. Instead of focusing on fractures and blood loss, they focused on the brain alone.

As for actual treatments, they discussed trepanning the king's skull to remove any excess fluids and "corrupted" blood, but the risks outweighed the benefits, and they gave the idea up. In the meantime they examined the heads of the decapitated criminals. History doesn't record the exact methodology here—whether someone fixed each head inside a vise to provide a stable target, or perhaps strung the noggins up like piñatas to swing at—but Montgomery's lance stump got quite a workout battering their mugs. It was a macabre mix of medieval brutality and modern experimental savvy, and Paré and Vesalius eagerly examined the targets for clues. Alas, the heads offered little inspiration for treatment.

The two men could have learned a lot more by simply observing the king, whose suffering foreshadowed many great discoveries over the next four centuries of neuroscience. Henri continued to drift in and out of coherence, limning the borders of the unconscious. He suffered from seizures and temporary paralysis, two then-mysterious afflictions. Strangely, the paralysis or seizures would derange only half of his body at any one time, a clear hint (in retrospect) that the brain controls the body's halves independently. Henri's vision also went in and out, a clue that the back of the brain (where Paré expected to find the contrecoup damage) controls our sense of sight. Worst of all, Henri's headache kept widening, which told Paré that his brain was swelling and that blood vessels inside the skull had ruptured. As we know today, inflammation and fluid pressure can crush brain cells, destroying the switches and circuits that run the body and mind. This explains why brain injuries can be lethal even if the skull suffers no fracture. Skull fractures can in fact save people's lives, by giving the swollen brain or pools of blood room to expand into. The history of neuroscience has proved the brain amazingly resilient, but one thing it cannot stand is pressure, and the secondary effects of trauma, like swelling, often prove more deadly than the initial blow.

King Henri II of France finally succumbed to an intracranial hemorrhage at 1 p.m. on July 10. Queen Catherine ordered every church to say six requiem masses daily and ordered all church bells—which had been bleating for the king—silenced. Amidst this sudden, sinister quiet, Vesalius and Paré began their famous autopsy.

To cut open a king—to even suggest such a thing—was bold. In that era, anatomists might open someone up for one of two reasons, a public lecture or an autopsy. Both activities had a stink of the disreputable about them. By the mid-1500s a few cities, especially in Italy, had relaxed the old prohibition on dissections for teaching purposes, but only barely: authorities might allow one per year (usually in winter, to prevent spoilage), and then only of criminals, since an official sentence of "death and dissection" would wring a little more posthumous punishment out of the rogue. Most kingdoms limited autopsies to suspected cases of poisoning, infanticide, or other heinous acts. And in some cases an "autopsy" did not require actually cutting open the body. Why Catherine gave in to Paré and Vesalius and permitted a full, invasive autopsy of Henri isn't clear, since everyone knew who had killed him and how, but history remains grateful she did.

Vesalius had laid out the proper steps for opening the skull in *Fabrica.* This usually involved lopping the head off to make examining the brain easier, but out of deference to the king, he merely elevated the chin in this case, by placing a wooden block beneath the royal nape. Someone grabbed a fistful of the king's graying hair to steady the skull, while someone else (presumably Vesalius, the expert dissector) began sawing an inch above the eyebrows. After circling the head and removing the skull vault, he encountered the thin membranes (the meninges) surrounding the brain. In *Fabrica* Vesalius suggested that students nick the meninges with their thumbnails and unwrap them. He then encouraged students to plunge their fingers in and squeeze and fondle every fold: dissection was as much a tactile as a visual pleasure for him. But with Henri, Vesalius restrained himself once again—probably in part because Henri's brain didn't look all that appetizing. The front and sides looked normal, but in the rear—antipodal to the blow*—Vesalius and Paré found pools of blackened fluids beneath the meninges, like blisters about to burst. The brain itself had also yellowed and putrefied back there, a puslike mass that measured one thumb's width across by two thumbs' widths high.

Equally important, they found that the wooden shards from Montgomery's lance had never penetrated the brain.

It's not always clear what Vesalius and Paré understood, in modern terms, of how brain damage kills. In their reports they often lapsed into talk of imbalanced humors and "animal spirits" escaping Henri's body. They knew nothing of neurons or localization. And the shards from Montgomery's lance probably led to an infection that weakened Henri and hastened his death—a complication they couldn't have grasped. But the duo understood well enough that the "commotion" and "corruption" in the back of Henri's brain, along with the resultant pooling of blood, had ultimately killed him. Trauma to the brain alone, they determined, could be deadly, even without a skull fracture. And in proving this, they vastly outdid the mutterings of that old phony, Nostradamus. Nostradamus had bloviated about lions and cages of gold. Vesalius and Paré had predicted what sort of damage they'd find inside Henri's brain and exactly where they'd find it—and find it they did. They proved science the superior clairvoyant.

~

The fallout from Henri's death poisoned most everything he loved. After him, French kings were forbidden from jousting, for their own protection. Diane de Poitiers had to surrender the jewels and estates and place at court she'd earned as Henri's mistress. The new French king, the frail François II, died just seventeen months later, after contracting an earache while hunting. The next king in line, Charles IX, was ten years old, so Catherine assumed power as regent—putting an Italian, a Medici, in charge of France.

Henri's death had actually crushed Catherine: despite his shabby treatment of her, she loved him. (She even swapped her original royal symbol, a rainbow, for a broken lance.) But her policies over the next few years betrayed his hopes for peace and precipitated decades of civil war between Catholic Royalists and Protestants. These wars reached their nadir with the St. Bartholomew's Day Massacre in

August 1572, which Catherine likely engineered. Although intended as a surgical strike against key Protestant leaders, the killing fed on itself, and mobs spread across the countryside, butchering thousands; historians call it less a day of massacre than a season. One Protestant targeted was none other than Gabriel Montgomery, who, while in exile after manslaughtering Henri, had renounced Catholicism. After the St. Bart's massacre Montgomery fled to England, but he returned the next year to battle the royalists, capturing Normandy and threatening to conquer all of northern France. A lengthy pursuit ended with royalist troops capturing him in 1574, and Catherine had the pleasure of seeing the man she still blamed for her husband's death quartered and then beheaded.

As for the scientists, Paré had treated François II on his deathbed in 1560. The boy's earache had led to a buildup of fluid on the brain, but once again Paré declined to trepan a king of France. No one quite knows why he refused, and nasty rumors have always circulated that Paré (à la *Hamlet*) slipped poison into the young king's ear, probably at Catherine's request, so she could reign as regent. But there's another reason Paré did not perform emergency neurosurgery. The risks involved with trepanning were high, and he knew he would likely incur the blame for any mishap. That was doubly true since Paré had converted to Protestantism by this time, and therefore—far from being someone that Catherine would entrust with a murder—actually held a precarious position in Her Majesty's government. Indeed, Paré barely survived the St. Bartholomew's Day Massacre a dozen years later.

Nevertheless, during the intervals of peace in Paris, Paré thrived. He wrote a handbook for military surgeons and an anatomy textbook that plagiarized Vesalius. (Paré didn't see this as a big deal, calling his appropriation "as harmless as the lighting of one candle from the flame of another.") He also campaigned against mummies and unicorn horns and other bogus cures. Most important, Henri's autopsy inspired Paré to write a book about head wounds. It called attention to the danger of contrecoup injuries and pooling fluids, and it continued the vital work of pairing specific brain injuries with specific symptoms—the modus operandi of neuroscience for the next four centuries. The world's finest surgeon spent his twilight in Paris, having served four kings, and he died in his bed in one of his five houses.

Vesalius met a nastier end. Within a month of Henri's autopsy, King Philip quit cold Brussels for sunny Spain. Vesalius followed, and soon wished he hadn't. There are two competing stories about what finally drove Vesalius from Spain. The less likely story says that Vesalius got a little too anxious to start the autopsy of a noblewoman one night—and found her heart still beating when he cut her open. Her family supposedly called in the Inquisition, and Vesalius saved his neck only by agreeing to make a pilgrimage to Jerusalem.

The second story, while probably more truthful, is even stranger. The Spanish heir apparent, Don Carlos, called the Infante, was a weak and febrile boy. No one had much sympathy for him, however, since he was also a psychopath. Born with teeth, he delighted in gnashing his nursemaids' nipples until they bled and became infected, and he spent much of his childhood roasting animals alive. By his teenage years he'd moved on to deflowering young girls. One night in 1562 the Infante tore down the stairs to snatch a maiden he'd spied, but karma tripped him. He somersaulted and smashed his noggin at the bottom of the staircase, lying there bleeding for some time. Spanish doctors failed to cure the prince, so Philip sent Vesalius. Vesalius found a tiny but deep red wound at the base of the prince's skull, and he suggested trepanation to alleviate pressure. The Spanish doctors, spiteful at a foreigner's interference, refused. Instead, they allowed the local townsfolk to dig up the desiccated, century-old corpse of Friar Diego, a cook at a local monastery and a reputed miracle-worker. The townsfolk then entered the Infante's bedchamber to slip Diego beneath the boy's sheets—and the boy, who was more or less out of his wits by then, snuggled up to and began dreaming of visits from the friar. A few days later he'd improved little, and Vesalius finally prevailed upon the other doctors to puncture the skull near the eye socket and drain some pus. The Infante recovered within a week after this, but the doctors and townsfolk universally credited Diego, who was later canonized for Vesalius's miracle.

The whole farce disgusted Vesalius and convinced him to quit Spain. So he arranged a holy pilgrimage to escape. He first visited Padua, where he'd produced *Fabrica,* and arranged to get his old job as professor back. Nevertheless, perhaps feeling guilty about using a pilgrimage as a ruse, Vesalius continued to the Holy Land, landing at Jaffa in the summer of 1564. He visited Jerusalem and the plains of Jericho, and sailed back satisfied, but he never reached Padua. He'd booked passage on a cut-rate tourist ship with inadequate supplies, and when storms ravaged the vessel on the return voyage, passengers began expiring from a lack of victuals and fresh water. Like something out of Géricault's *The Raft of the Medusa,* corpses were being heaved overboard, and for once in his life the sight of dead bodies spooked Andreas Vesalius. He went half mad, and scrambled ashore as soon as the ship staggered to Zakynthos, an island in what's now western Greece. According to different accounts, he either died at the gates of Zante, a port city, or crawled to a filthy inn where the locals, wary of the plague, let him die alone. Either way, it was an anticlimactic death. There was no autopsy to determine what had killed him.

In the end, about the only person, place, or thing to gain from Henri's death was the incipient field of neuroscience. On a basic level, Henri's autopsy confirmed beyond a doubt that contrecoup injuries existed, and that the brain could suffer trauma even if the skull remained undamaged. It's a lesson, sadly, we're still relearning today. Rope-a-dope boxers and quarterbacks and hockey enforcers continue to shake off concussions on the theory of *no blood, no harm.* But each concussion effectively softens up the brain and ups the chances of more concussions. After multiple blows, neurons start to die and spongy holes open up; people's personalities then disintegrate, leaving them depressed, diminished, suicidal. Four centuries have passed, but macho modern athletes* might as well trade pads for armor and go joust with Henri.

On a deeper level, Henri's death helped inaugurate a new approach to neuroscience. You can't call Vesalius and Paré modern: each revered Galen along with Hippocrates and the rest of the Greek medical chorus. But each of them also evolved past the ancients, by emphasizing experiments and observation. Vesalius bequeathed a new map of the brain, Paré new diagnoses and surgical techniques; and while Henri's was not the first autopsy, in terms of prestige—prestige of both the patient and the practitioners—it was the summa of early medical science. The treatment of royals often defined what became standard care for everyone else, and after Henri's death, autopsies started to spread throughout Europe. This expansion made it easier to correlate specific brain damage with altered behavior, and with every new autopsy, neuroscientists learned to pinpoint people's symptoms more precisely.

Soon scientists even moved beyond the brain's gross anatomy, into a realm that Paré and Vesalius never dreamed of, the microscopic. Like physicists drilling down into the fundamental particles of the universe, neuroscientists began to drill down, down, down into the fundamental matter of the brain, parsing it into tissues and cells and axons and synapses before finally arriving at the brain's basic currency, its neurotransmitters.

PART II

CELLS, SENSES, CIRCUITS

CHAPTER TWO

The Assassin's Soup

Now that we've gotten an overview of the brain, we'll investigate it piece by piece in upcoming chapters, starting with its smallest bits—the neurotransmitters that relay signals between cells.

God's ways are not man's ways, God's reasons not man's reasons, so when God told Charles Guiteau to shoot the president, Charles Guiteau agreed. And if doing so simultaneously saved his beloved Republican Party, so much the better.

God and Guiteau (pronounced *Git-OH*) went way back. During Guiteau's childhood, his mother used to shave her head and lock herself into her bedroom to chant Bible passages. His father obsessed over the millenarian sermons of one John Noyes, and after failing a college entrance exam, Charles himself joined Noyes's sex-crazed utopian cult in Oneida, New York. Guiteau whiled away the Civil War there, but even the polyamorous, free-lovin' ladies at Oneida froze him out, repelled by his bulging eyes, lopsided smile, and monomania. They mocked him as "Charles Get-out."

After he got out, in 1865, he began evangelizing—first founding a newspaper, *The Daily Theocrat,* which flopped, then trying his hand at preaching, charming crowds with talks like "Why Two-Thirds of the Human Race Are Going Down to Perdition." He also self-published a book, *The Truth,* about Christ's second coming. Much of the book was mad—he called Stanley and Livingstone signs of the apocalypse—and what wasn't mad he plagiarized from Noyes. Along the way Guiteau passed the bar exam (depending on the year, it contained three or four questions; he needed to get only two right), but lost his first case after frightening the jury with a fist-shaking, spit-flecked rant. He started a debt-collection service but mostly pocketed his clients' cash, and after running out of boardinghouses to get blacklisted at, he moved to Chicago to mooch off his sister, Frances, and her husband, a lawyer named George Scoville. That cozy arrangement ended when he swung an axe at Frances. He drifted back to New York, where he married a YMCA librarian whom he punched and kicked and locked into closets for sassing back. She divorced him, but only after nursing him back to health after he caught syphilis from a hooker. The disease eventually infected his brain.

Naturally, Guiteau thought himself fit for politics. A stalwart Republican, he wrote a clichéd stump speech in 1880 supporting Ulysses S. Grant's bid for a third term as president. When James Garfield got the Republican nomination instead, Guiteau simply swapped in Garfield's name. He then begged Garfield's campaign crew in New York—including Chester Arthur, nominee for vice president—for chances to deliver it. The party finally sent him to a rally for black laborers. Struck with stage fright, Guiteau muttered a few paragraphs and quit. He nonetheless convinced himself that he'd delivered New York for Garfield. So after Garfield won the election, Guiteau spent some of his last dollars on a train to Washington, D.C., to claim a job in the new administration.

Him and about a million others. This was the peak of the spoils system, which transformed the first months of any presidency into a job fair. Despite not speaking a foreign language or ever having traveled abroad, Guiteau decided to seek a post in Europe. After waiting in line for hours, he finally met Garfield and handed him the speech that "clinched" New York, with "Paris consulship" scrawled across the top. By this point Guiteau was down to his last shirt; he wore rubber rain covers for shoes and owned no socks. But he smiled his best zigzag smile at Garfield and withdrew, letting the president puzzle over what the hell had just happened.

In those days common citizens could visit the White House unannounced, and in late March, Guiteau began nagging Garfield's secretaries and even Cabinet members for news about his Paris post. The secretary of state finally screamed at Guiteau to shut up already, and when Guiteau got caught nicking White House stationery, he was banned. Guiteau nevertheless—the man was a true, true optimist—kept scanning the newspapers for word of his appointment.

It never came. And other notices in the papers disturbed him even more. Garfield—despite previous success as a college president, Civil

War officer, and congressman from Ohio—soon found his administration floundering. A few broken promises had caused a rift among Republicans, and both Republican senators from New York resigned in a snit. With every damning headline, Guiteau's buggy eyes bulged even more: the GOP was disintegrating. Someone had to save it.

Kill Garfield. God first whispered this to Guiteau in May of 1881. Although stunned that, as he put it, "Jesus Christ & Co." had selected him for the deed, the more Guiteau pondered it, the more logical it seemed. *Kill Garfield.* Yes, with Garfield gone, his New York chum Chester Arthur would assume power and calm the Republican waters. Arthur would then of course pardon Guiteau once Guiteau explained about God's instructions. Heck, he might see Paris yet.

Guiteau borrowed ten dollars and bought a "British Bulldog" revolver at a gun shop one block from the White House—paying extra for an ivory handle, since it would look handsomer in a museum someday. Guiteau had never fired a handgun, so he marched to the Potomac Tidal Basin for practice. The gun's recoil almost somersaulted him into the mud, and he hit his target just once—blowing a hole in a sapling. Ever confident, though, he began stalking the president that very week. He also began revising *The Truth,* sure to become a bestseller soon.

Guiteau decided to assassinate Garfield in church, and tailed him there one Sunday to do some recon. Despite the need to be inconspicuous, a revved-up Guiteau stood up at one point and shouted at the preacher, "What think ye of Christ?" (In his diary, Garfield recalled "a dull young man, with a loud voice.") Guiteau changed his mind later that week, and decided to shoot Garfield at the train station instead. But he backed down, all mushy, when he saw Mrs. Garfield walking arm in arm with her man.

Weeks later Guiteau called off a third assassination attempt because the weather got too hot, then a fourth because he didn't want to interrupt an important-looking conversation between Garfield and the secretary of state. At last the newspapers announced that Garfield would leave D.C. for the summer on July 2, and Guiteau steeled himself to act. On the big day he arose at 4 a.m., took some practice shots by the Potomac, got his boots polished, and rode a cab to the train depot, where he unwrapped his gun in the bathroom and waited.

Garfield woke up buoyant that morning, eager to abandon the stinking swamp of Washington and all its partisan spats and grubby job-seekers. He burst into the bedroom of his young boys, Abram and Irvin, and clowned around like a teenager—doing handstands, singing Gilbert and Sullivan, vaulting over the bed to prove the old man still had it. He arrived at the depot around 9:20 a.m., and walked with an advisor toward his train.

Kill Garfield. Slinking out, Guiteau closed to within two yards. The first shot grazed Garfield's arm, stunning him. Guiteau fired again and plugged Garfield in the lower back. This second shot brought pandemonium on the platform—screams, hollers, chaos. Guiteau speedwalked away, but a policeman nabbed him at the depot exit.

Meanwhile Garfield's legs crumpled and he sank down, a circle of red blooming on his back. Two doctors arrived moments later—as did Garfield's advisors, including Robert Todd Lincoln, who sixteen years before had seen his father borne out of Ford's Theatre. "Mr. President, are you badly hurt?" one doctor asked. According to one account, Garfield breathed, "I'm a dead man."

So began the national James Garfield deathwatch. With the recent expansion of telegraph lines across the world, Garfield's suffering became practically a live event, and Garfield's physician, one Dr. Doctor Bliss (*sic* the first name, and last), took full advantage of the new medium. Newspapers coast to coast reprinted his daily bulletins, and many cities plastered updates onto huge billboards in their public squares.

Unfortunately, Dr. Doctor provided more bliss with his public relations than with his medical care. Garfield suffered from three main problems over the next few months: isolation, hunger, and pain. Isolation, because Bliss confined him to a bed and forbade even family members from seeing him at first. Hunger, because Bliss, fearing an intestinal infection, began feeding the president rectally, with a slurry of beef broth, egg yolks, milk, whiskey, and opium. (The empty-bellied president spent many an hour that summer fantasizing about hearty recipes from his frontier boyhood such as squirrel soup.) Pain, because Guiteau's second shot had lodged itself inside Garfield's torso: he described the discomfort as a "tiger claw" raking his legs and genitals. Bliss tried excavating the bullet, but no matter how many times he worked his fingers into the wound and rooted around in Garfield's groin, the slug eluded him. Other doctors tried their hands, too, and Bliss even recruited Alexander Graham Bell to rig up a crude metal detector of batteries and wire. No trace. A few doctors begged Bliss to check near Garfield's spinal cord instead, since the collapse of the president's legs at the depot and the subsequent shooting pain both sounded like neurological trouble. Bliss blew them off and kept digging. Meanwhile, he kept releasing what one historian called "fraudulently optimistic" bulletins about Garfield's progress and sure recovery. Other doctors leaked more negative assessments, causing a rift within the president's medical team.

Bliss eventually granted Garfield's wish to escape D.C., and they relocated to the president's cabin in coastal New Jersey. Ironworkers laid 3,200 extra feet of track right to the cabin door, then pushed Garfield's railroad car the last quarter mile when it got stuck on a hill. The change of scenery and seaside air buoyed the president at first, but he soon faded, as he still couldn't eat. Overall, Garfield lost eighty pounds in eighty wretched days, and when Bliss's fingers finally infected Garfield's wound, turning it into a slimy pocket of pus, Garfield had no fight left. He died on September 19, 1881. The autopsy found the bullet nestled near his spine.

The public, both North and South, howled in anguish. Garfield had been the American ideal, a genuine rag-to-riches president, and the mourning for him—not to mention the loathing for Guiteau—united the country for probably the first time since before the Civil

War. In fact, Guiteau almost didn't make it to trial later that year, as two Jack Ruby wannabes tried to exact revenge. One (Guiteau's jailer) missed from five feet, and one put a bullet through his coat but missed everything vital.

Guiteau finally stood trial in November, with George Scoville, his poor brother-in-law, handling the defense. Overwhelmed—he normally handled land deeds—Scoville submitted an insanity plea. Guiteau scoffed, thinking himself perfectly sane—would God have chosen him otherwise? But he unwittingly made Scoville's case for him by repeatedly interrupting the trial: at various points, in his whiny, scratchy voice, he chanted epic verse, sang "John Brown's Body," called the jurors jackasses, and announced his own candidacy for president in 1884. He ranted, too, that the murder charge was unfair: he'd merely shot Garfield; *doctors* had killed him. (Here, he was probably right.) Nor were Guiteau's antics limited to courtroom hours. Newspapers caught him selling autographed glamour shots from his cell—nine dollars per dozen.

Amazingly, the insanity defense went nowhere, even after Guiteau compared himself to Napoleon, St. Paul, Martin Luther, and Cicero. The public's appetite for revenge had grown too keen, and the prosecution whetted it further by passing around Garfield's shattered vertebrae. Besides, psychiatrist after psychiatrist testified that Guiteau understood right and wrong, and was therefore sane. Of 140 witnesses, in fact, only one man maintained, unshakably, without qualification, that Guiteau had lost his mind.

Though just twenty-nine, Edward Charles Spitzka had already earned renown in certain circles as a brain pathologist. Guiteau's case made Spitzka famous, period, in part because he testified despite receiving death threats from citizens angry that he might get Guiteau off. Besides all the psychological signs of insanity, like the god delusion, Spitzka pointed out signs of neurological trouble in Guiteau as well. In particular, Guiteau's lopsided smile, lazy left eye, and lolling tongue implied that he couldn't control both sides of his face equally well. In all, Spitzka would later call Guiteau the most "consistent record of insane manner, insane behavior, and insane language...in the history of forensic psychology."

The jury disagreed and found Guiteau guilty in January 1882. Remanded to his cell, Guiteau waited months for a pardon from Arthur. When it didn't come, he shrugged, eager to taste the fruits of paradise. On the scaffold, near the Anacostia River, he even recited a poem he'd written for the occasion: "I'm Going to the Lordy." (The city denied his request for orchestral accompaniment.) As the hangman hooded him, blotting out his crooked smile for the last time, Guiteau let the verses waft from his hands. A moment later he fell earthward himself.

The autopsy took place ninety minutes later, at 2:30 p.m. Overall Guiteau's body looked sound, aside from the rope burns on his neck; Guiteau had even (like most hanging victims) gotten an erection and ejaculated before dying. The bigger question was whether Guiteau's brain looked sound. Most scientists at the time believed that insanity,

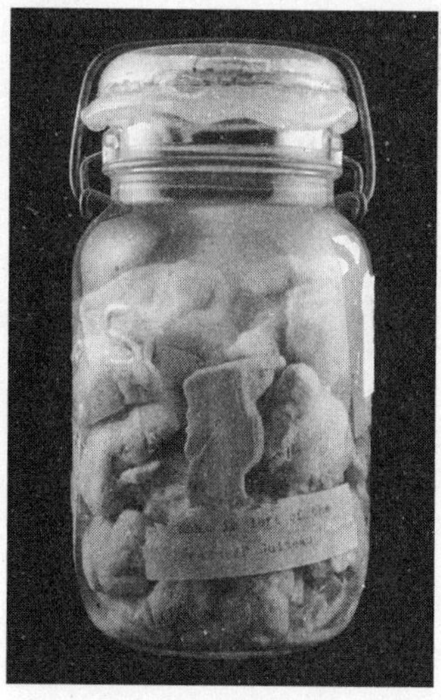

Assassin Charles Guiteau (left) and his brain (right). The caption on the jar reads: "What is left of the brain if *[sic]* Guiteau." (National Library of Medicine)

true insanity, always betrayed itself by clear brain damage—lesions, hemorrhages, putrid tissue, something. Inside Guiteau's skull, though, nothing seemed amiss at first. His brain weighed fifty ounces on a grocer's scale, a tad more than average, and beyond some small anomalies (extra creases here and there, a slightly flattened right hemisphere) his brain looked normal, chillingly normal.

Since the days of Vesalius and Paré, though, autopsies had become much more of a microscopic art. And under a microscope, Guiteau's brain looked awful. The outer rind on the surface, the "gray matter" that controls higher thinking, had thinned to almost nothing in spots. Neurons had perished in droves, leaving tiny holes, as if someone had carbonated the tissue. Yellow-brown gunk, a remnant of dying blood vessels, was smeared everywhere as well. Overall the pathologists found "decided chronic disease... pervad[ing] all portions of the brain." As Spitzka had testified, Guiteau was surely insane.

Still, because the stigmata of insanity—the physical signs of brain damage—appeared only on the microscopic level, most neuroscientists continued to dispute the evidence, because most neuroscientists at the time didn't appreciate the importance of microanatomy. Only over the next two decades, in fact, would neuroscientists take the first real steps toward explaining how brain cells work. This understanding would emerge just in time for the next assassination of a U.S. president and the next national dispute over criminal insanity.

~

By the later 1800s most biologists believed in "cell theory"—the idea that living creatures are composed entirely of tiny, squishy bricks called cells. Neuroscientists weren't so sure. Discrete cells might exist in the rest of the body, yes. But under the microscope, neurons seemed to have no breaks or gaps between them; instead, they seemed fused together into one large, lacy net. Moreover, neuroscientists believed that—unlike other, more autonomous cells—neurons *acted* in unison as well, pulsing and thinking as one. They called this big net of neurons the neural reticulum.

The undoing of the neural reticulum theory started with an accident one night in 1873. According to legend, Camillo Golgi was working by candlelight in the kitchen of an old Italian insane asylum when he elbowed over a beaker of silver solution onto some slices of owl brain. *Merda.* The silver solution was used to stain tissues, and Golgi assumed that his clumsiness had ruined his samples. Still, he examined them under his microscope a few weeks later—and found, to his delight, that the silver had stained the brain cells in a peculiar, and most useful, way. Few cells absorbed the silver overall; but those that had absorbed it stood out dramatically—black silhouettes on a sherbet-yellow background, with their finest fibers and tendrils suddenly visible. Fascinated, Golgi set about refining this staining technique, which he called *la reazione nera,* the black reaction.*

Scientists at the time already knew that the nervous system contained two main types of cells, neurons and glia. (Neurons process thoughts and sensations in the brain, and also make up the bodies of nerves. Glia, meaning "glue," hold neurons in place and provide nutrients, among other jobs.) Golgi, however, became the first person to see these cells in anything like full detail. The rounded glia with their wispy tentacles astounded him, like black jellyfish frozen in amber. The neurons looked equally exotic, being composed of three distinct parts. Each neuron had a circular central body; an intricate bush of "dendrite" branches sprouting from the body; and a glorious "axon," an arm that spooled out incredibly far from the body, twisting and turning for microscopic miles before erupting into its own tiny branches at the far tip. Golgi deduced that neurons must communicate via their axons, since the branches at the far tip were often entangled with other neurons. In fact, the axons were so closely entangled that Golgi could see no space between neurons, and he came out as a strong supporter of reticulum theory.

Other neuroscientists, including Santiago Ramón y Cajal, found *la reazione nera* every bit as enchanting as Golgi did. "[It] renders anatomical analysis both a joy and a pleasure," Cajal gushed, and he compared the stains to "fine India ink drawings on transparent Japanese vellum." That's awfully specific, but Cajal would know: he'd aspired to be an artist while growing up in Spain. That dream ended at age ten, however, when a local landscape painter declared Cajal talentless, prompting the boy's father to seize his brushes and easels and enroll him in a Jesuit school. Bored and angry, Cajal began lashing out, and at age eleven was jailed for building a cannon from an oil drum and blowing down a neighbor's gate. Cajal's father tolerated this, but when the boy's grades started slipping, he yanked him from school and apprenticed him to a barber. Suddenly appreciative of his education, Cajal reenrolled in school and began studying various topics in medicine, including hypnosis. He settled on neuroscience, and Golgi's reaction opened his eyes to the beauty of the field, allowing him to fuse neuroscience with art.

However much he adored Golgi's artistry, though, Cajal disagreed with Golgi's conclusions, especially about the brain's gray matter. Anatomically, the brain contains two distinct substances, gray matter and white matter. Gray matter has a high percentage of neurons, and most gray matter resides on the brain's surface, in a wrinkly rind called the cortex. (Or at least most gray matter resides *near* the surface: two-thirds of the cortex is actually invisible from the outside, being corrugated and folded up just beneath the surface. If unfolded and smoothed out, the cortex would be the rough size of a pillowcase, but would be only one-tenth of an inch thick.) After inspecting hundreds of stains with his microscope, Cajal saw that the gray matter didn't look at all like what Golgi claimed, with all the neurons fused together. Cajal saw discrete neurons. Moreover, when Cajal strangled a few neurons as an experiment and let them wither, the decay always stopped at the border of the next neuron instead of killing it, too, as you'd expect if both were really fused together.

Cajal also rejected Golgi's metaphor for the macro organization of brain cells: rather than a horizontally spread neural hairnet, Cajal saw neurons arranged into tiny vertical "columns" of around one hundred neurons each—little stacks that covered the surface of the brain like stubble. Axons from one column did sometimes reach horizontally into neighboring columns, Cajal admitted, but vertical organization* was the general rule.

Finally, while Golgi believed that neurons communicated exclusively via their axons, Cajal ruled otherwise. Near the eyes, for instance, Cajal saw *dendrites* turned toward the retina, poised to absorb information. And within long chains, neurons usually lined themselves up axon to dendrite, one after the other. Indeed, the axons of one neuron "fit" into the dendrites of the next neuron like a hundred-fingered hand sliding into a hundred-fingered glove. All this could mean only one thing: neurons might speak with axons, but they listened with dendrites. Both were essential for communication.

These findings led Cajal to propose the "neuron doctrine," one of the most important discoveries ever in neuroscience. In brief, Cajal's neurons were not continuous, but had tiny gaps between them. And they transmitted information in one direction only: from dendrite to cell

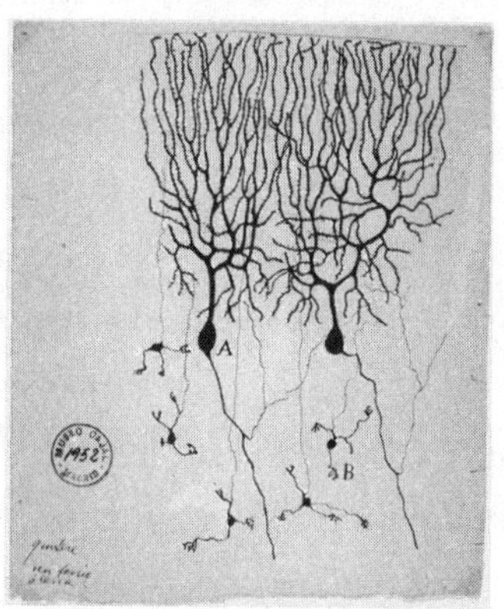

Left: Gorgeously intricate neurons drawn by Santiago Ramón y Cajal, a neuroscientist and sometime artist. Right: Cajal himself.

body to axon. That is, no matter the signal *(food! tiger! hubba-hubba!)*, it always entered a neuron through its dendrites; passed through the cell body for processing; and only then shunted down the axon. (I think of this progression, reverse-alphabetically, as d → cb → a.) When the signal reached the axon tip, the neuron then tickled the dendrites of the next neuron in line, and the process started over. Golgi might have seen the true shape of neurons first, but Cajal first determined how those silhouettes worked.

Still, Cajal found the neuron doctrine a tough sell. He actually had to launch his own journal to propagate his ideas, and even this didn't help, since few medical scientists bothered reading Spanish journals. So in 1889 he staked his career on traveling to a conference in Germany, the world's greatest scientific center, even paying his own way when his university refused. Luckily for him, his gorgeous freehand drawings of neurons won him some converts. And over the next decade the neuron doctrine gained a foothold in scientific circles—albeit grudgingly. Many scientists still refused to believe Cajal, and by 1900, two armies of neuroscientists had drawn up lines, with Golgi's "reticulists" and Cajal's "neuronists" despising each other more and more each year.

History loves a good gag, though, so it was inevitable that Golgi and Cajal would share a Nobel Prize in 1906. Cajal groaned about this, lamenting the "cruel irony of fate, to pair, like Siamese twins united at the shoulders, scientific adversaries of such contrasting character." In their acceptance speeches both men, especially Golgi, kept bashing each other's "odious errors" and "deliberate omissions." The Peace Prize this wasn't.

Ultimately, the neuron doctrine triumphed because it explained so much more. Even the unraveling of Charles Guiteau's mind began to make sense. Guiteau's autopsy found extensive damage to his glia, which support and nourish neurons. Minus that support, neurons languished and died—especially in his gray matter, which was reduced to a combover in spots. And even where neurons had survived, they often had fewer axon and dendrite branches than normal, further reducing their ability to communicate and process thoughts. In retrospect, Guiteau's brain offered ugly but definitive proof of how neurons work, or fail to.

Like all great discoveries, though, Cajal's neuron doctrine opened up as many new questions as it answered. Most important: if neurons were separate, how exactly did a signal cross the gap between them? There seemed just two possibilities, electric currents or chemical pulses. Once again, each side in this battle had its partisans and stalwarts, with the "sparks" championing electricity and the "soups" championing a mélange of biochemicals. And once again, the dispute between them would bleed beyond the realm of neuroscience and color the debate over the sanity of an enigmatic assassin.

~

Smile pump nudge. Smile pump nudge. Smile pump wink laugh smile pump nudge—on September 6, 1901, William McKinley was *on.* Throughout his five years as president, McKinley had adored mingling with crowds: flirting with housewives, tipping his hat to bankers, pinching the cheeks of beribboned girls. But to keep the lines of well-wishers from clotting, he'd developed the "McKinley handshake." He'd smile broadly and intercept the person's fingers above his palm, so he could disengage at will. He'd then cradle their elbows with his left hand and gently nudge them off balance, moving them aside and clearing room for the next mark. Smile pump nudge—fifty handshakes per minute. But while visiting Buffalo on September 6, a mustachioed foreigner outmaneuvered McKinley. He clasped the presidential palm and, even as one of McKinley's guards lurched forward, held the shake suspiciously long.

The Pan-American Exposition in Buffalo had been wowing crowds for months, with bullfighting and mock Japanese villages and fountains fit for Versailles. The Trip to the Moon exhibit featured midgets dressed like aliens and serving green cheese, and the 389-foot Electric

Tower—a spire illuminated with thousands of "electric candles" (i.e., lightbulbs)—shone so beautifully at night that people wept. McKinley's four-day visit crowned the expo as the national event of the year, and McKinley responded by giving, on the afternoon of September 5, the finest speech of his presidency, about the boundless prosperity of the United States.

Amid the cheering crowd of fifty thousand, though, one man—a laborer with a slight build, a thin mustache, and little hope of sharing in that prosperity—seethed. Leon Czolgosz had started working full-time at age ten, in 1883, and by 1893 was making four dollars a day spinning wire near Cleveland. But his mill cut wages during the 1893 financial panic and fired him when he joined a strike. A committed Republican before this, Czolgosz (*chol-gawsh*—think *z* = *h*) declared himself a socialist. The burgeoning socialist movement had clashed repeatedly with factory owners that decade, and the conflict had split the country ideologically. The brutal conditions in factories appalled most people, but Americans felt equally scared of the unruly mobs marching in the streets, rioting and spouting off about revolution.

Czolgosz eventually got his job back under a false name. But his working career ended when he had a mysterious mental breakdown in 1898. He retreated to the family farm, where he loafed away most afternoons, occasionally doing mechanic jobs but mostly hunting rabbits and leafing through socialist tracts. He also became withdrawn, taking meals of raw milk and crackers alone in his room in the attic—possibly because he feared his stepmom, Catrina, wanted to poison him. (Shades of paranoia.) His one happy memory from these years came in July 1900, when he read in the newspaper that an Italian American silk weaver, Gaetano Bresci, had assassinated King Umberto I of Italy. Bresci's bravery riveted Czolgosz, who clipped and kept the story.

In May 1901, Czolgosz heard the anarchist leader Emma Goldman speak in Cleveland. Goldman sometimes glorified assassinations, and as Czolgosz later told his jailers, when he heard Goldman, "My head nearly split... She set me on fire." He converted from socialism to anarchism on the spot. He then tailed Goldman to Chicago, where he badgered local anarchist leaders, calling them "comrade" and asking in conspiratorial tones to attend their "secret meetings." Overall, most anarchists found Czolgosz pathetic. Others considered him either ignorant (he didn't seem to grasp the contradictions between socialism and anarchism, for one thing) or outright dangerous: the editor of an anarchist newspaper denounced him, in print, as a police narc.

To prove himself, and "to do something heroic for the cause," as he said, Czolgosz took a room above a saloon in Buffalo on August 31, telling people he planned to sell souvenirs at the expo. The proprietor remembered liking Czolgosz, since he paid his two dollars rent up front and drank good whiskey, not the nickel-per-shot turpentine that most patrons did. Sometime that week Czolgosz bought the same make of revolver that Bresci had used to kill King Umberto, a $4.50 silver-plated "Saturday Night Special." A good hunter, Czolgosz needed no target practice, but he did spend hours alone at night (shades of Travis Bickle) yanking the gun from his pocket and quick-wrapping it in a white handkerchief to conceal its glint.

Czolgosz met McKinley's train when it arrived in Buffalo on September 3. Before he could shoot, though, a fusillade of cannons—there to welcome the president—fired instead. The concussion shattered windows on the train and put McKinley's security detail on edge, so Czolgosz slunk off. For the next three days he stalked "the ruler" (his words) around the expo, shadowing him near the Pylon of Liberty and Streets of Mexico bullfighting ring and other exhibits. But Czolgosz never could get a clean shot.

On September 6, the president's last day in Buffalo, McKinley toured Niagara Falls, which powered the electric dynamos that lit the expo. Reporters recorded a hairy moment that morning, when McKinley's coach drew near a chalk line marking the international boundary with Canada. No sitting president had ever left the country before, and McKinley warned his coachman to give the line a wide berth.

That crisis averted, McKinley enjoyed a buffet lunch at a hotel. He then said goodbye for the day to his wife, who was tired and flushed with heat. With only the menfolk left, they broke out the cigars and shot the bull. One local bigwig remarked that McKinley sure seemed to enjoy Buffalo. McKinley teased, "I don't know if I'll ever get away."

By midafternoon McKinley had one event left, a ten-minute meet-and-greet at the rococo Temple of Music, a 22,000-square-foot terra-cotta dome with garish pastel trim. People had started lining up hours before, dabbing their faces with handkerchiefs in the eighty-two-degree heat. Near the front of the queue, to pass the time, a six-foot-four-inch black waiter named James Parker tried to chat up a freshly shaven young man. Leon Czolgosz snubbed him.

At 4 p.m. McKinley's guards wrenched open the temple's immense doors and channeled the crowd into an aisle of chairs draped with bunting. Off to one side an organist played Bach on one of the largest pipe organs in the country. Czolgosz could just make out McKinley at the head of the line: "the ruler" stood amid a jungle of potted trees, beneath two gigantic U.S. flags. Smile pump nudge, smile pump nudge. McKinley paused only once, to give away the lucky red carnation in his buttonhole to a young lass.

Around 4:07, one of McKinley's guards noticed a swarthy, mustachioed Italian man who seemed a little too eager to meet the president. The guard considered intercepting him, but hesitated. At that moment the Italian grabbed McKinley's palm and pulled the president close. The guard's gut clenched, and he jumped forward. He broke up the tête-à-tête, then swiveled to watch the suspect depart.

Meanwhile, a man with his right hand bandaged in a handkerchief stepped forward. Czolgosz was so close that his first shot left powder burns on McKinley's vest. The puny bullet struck a button, however, and caromed off McKinley's sternum; doctors later found the bullet wrapped inside his clothes. The second shot struck home, tearing a hole in McKinley's stomach and pancreas. The handkerchief, still wrapped around the gun, caught fire.

Czolgosz had wanted, like Bresci, to squeeze off five rounds, but big James Parker, the waiter behind him in line, clobbered Czolgosz's fist, then smashed him in the face. Another guard swooped in, then ten more, and Czolgosz fell to the floor in a blur of boots and rifle butts.* A few feet away guards led McKinley to a chair, blood overflowing his waistband. After a few breaths McKinley noticed the scrum around Czolgosz and called out, "Go easy on him, boys." (This probably saved Czolgosz's life.) Moments later McKinley's entourage of advisors converged on him—including Robert Todd Lincoln, that Hope Diamond of nineteenth-century Republican presidents.

The expo ambulance—an electric "horseless carriage," one of the first electric cars—spirited McKinley off to a glorified first-aid station nearby. Buffalo's best surgeon was elbow deep in another patient just then, so officials grabbed the most senior medical man they could find, a gynecologist. His hair half cut—he'd been yanked from the barbershop—the gynecologist prepped for surgery while McKinley huffed ether. Frustratingly, despite the electric lighting elsewhere at the expo, the clinic wasn't wired. Even when assistants used mirrors to reflect the fading sunlight into the wound, the doctor couldn't see much. He managed to patch the stomach but couldn't find the second bullet, and he sewed McKinley up without draining the wound.

Meanwhile, thousands of people swarmed the Temple of Music, howling to lynch the assassin; some brandished ropes they'd torn off nearby exhibits. Buffalo's finest barely got Czolgosz to jail alive. A search of his person revealed, among other effects, $1.54, a pencil, a rubber nipple from a baby's bottle, and, legend has it, the newspaper clipping of the Bresci assassination.

In the week after his surgery, McKinley convalesced at the expo president's mansion. Teddy Roosevelt, McKinley's rambunctious vice president, rushed to his bedside. So did McKinley's wife, Ida; she'd suffered from epilepsy for years, and now reciprocated for all the hours that he'd nursed her. As with Garfield, his doctors fed McKinley rectally, and briefed the press daily about his temperature (approximately

102 degrees) and pulse (approximately 120). Although high, those numbers remained steady. And because McKinley remained coherent, even asking after Czolgosz once, people were confident that he'd recover. Roosevelt, in fact, soon skipped town for a hunting trip. McKinley's doctors also declined an offer to use one expo exhibit, a Thomas Edison–designed X-ray machine, to locate the second bullet. A *New York Times* headline on September 11 proclaimed, "President Will Get Well Soon."

McKinley nibbled his first solid food on the twelfth, toast and soft-boiled eggs. It was also his last solid food. His stomach and pancreas hadn't quite healed, and an infection roared to life inside him. By that night he was drifting in and out of consciousness. Aides tried frantically to locate Roosevelt, but he'd gone off the grid, deep in the Adirondacks. A park ranger finally spotted T.R. on September 13, and they tramped down a mountain in a midnight drizzle to catch a train to Buffalo. They were too late. McKinley had slid quickly, and died at 2:15 a.m. on September 14.

McKinley's death inflamed an already healthy public hatred of anarchists and immigrants. (Czolgosz was a U.S. citizen, Detroit born, but most decent folk sided with the *Journal of the American Medical Association,* which took one look at those diphthongs and spat, "thank God [he] bears a name that cannot be mistaken for that of an American.") Despite the national uproar, Czolgosz himself seemed indifferent to his fate: guards remembered a bicycle thief in the cell next to him going to pieces over being caught, while Czolgosz sat there phlegmatically day after day. He also grew his beard out, making him look far more like a stereotypical anarchist than before. To complete the picture of dishevelment, his jailers made him wear the same bloody clothes and underwear every day until the trial. Not that Czolgosz had long to wait: his trial opened September 23, just nine days after McKinley died. What followed was a low moment in American jurisprudence.

The trial lasted about eight hours total, over two days. This included two hours for jury selection, during which time all twelve jurors admitted they'd pretty much made up their minds already. Czolgosz, citing his anarchist creed, denied the legitimacy of his court-appointed defenders and refused to speak to them. They'd hoped to plead insanity, but all the alienists who'd chatted with Czolgosz had already declared him free of paranoia and delusions. (Rather than dig into his background or motives, the psychiatrists had mostly inquired into his reading habits, or tried to catch him in lies about the shooting. Two psychiatrists couldn't get Czolgosz to speak one word in two hours. They declared him fit for trial anyway.) Without the insanity defense, Czolgosz's lawyers basically gave up and focused on defending themselves, for having to take this "repugnant" assignment. They called zero witnesses, and when the case closed, the jury returned after half an hour—most of which time they'd spent debating how long they had to wait, for appearances' sake, before condemning Czolgosz. Two days later, in keeping with the major theme of the expo—the marvels of electricity—a Buffalo judge sentenced Czolgosz to die in the electric chair at Auburn State Prison.

The nation's first-ever execution by electricity, in 1890, had also taken place in Auburn—and had been overseen by Edward Charles Spitzka, the alienist who'd insisted that Charles Guiteau was crazy. Things had not gone well. The prisoner fried but refused to die, and his burning hair and flesh stunk up the tiny execution room. Spitzka screamed to flip the switch again, but the electricians had to wait two whole minutes for the generator to recharge. (In their defense, earlier tests that involved electrocuting a horse had gone much smoother.)

By 1901 Auburn had worked out the kinks. Guards awoke Czolgosz around 5 a.m. on October 29 and gave him dark trousers slit up the side. Inside the death chamber, an electrician wired up a string of twenty-two lightbulbs to test the current; when they began beaming, he pronounced the chair ready. Czolgosz entered at 7:06 a.m. and took a seat on "Old Sparky," a rough-hewn wooden throne sitting on a rubber mat. He promptly condemned the government again. Meantime, guards plopped a sponge soaked with conductive salt water onto his head. Next came the metal helmet, then another electrode clamped onto his calf beneath the slit in his pants. Last came the leather mask, which kept his face in place. It also muffled his final words: "I am awfully sorry because I did not see my father [again]." The electrician waited until Czolgosz exhaled—gases expand when heated, and the less air in the lungs, the less unsightly moaning during the death throes—and snapped the switch. Czolgosz jerked, cracking his restraints. After a few pulses at 1,700 volts, a doctor could find no pulse in Czolgosz. Time of death, 7:15 a.m.*

His hair still wet, his lips still curled from the shock of the shock, Czolgosz was laid out on a nearby table for the autopsy. One doctor dissected the body, while the all-important brain autopsy, including the determination of sanity, fell to a second doctor—or rather, a wannabe doctor, a twenty-five-year-old medical student at Columbia University.

Why entrust this job to someone with no medical license? Well, he'd already published scores of articles about the brain, including work on whether high doses of electricity damaged brain tissue or altered its appearance, an important consideration here. (Peripheral nerves usually fried, he'd found, but beyond a few small hemorrhages, the brain itself suffered little.) He also claimed phrenological expertise, including the ability to link mental deficiencies with unusual anatomical features. What clinched the selection was his pedigree—for this was Edward Anthony Spitzka, son of the Edward Charles Spitzka who'd defended Guiteau. No other father and son doctors can boast of involvement in two such historic cases. And whereas Spitzka *père* had failed to convince the world of Guiteau's insanity, Spitzka *fils* could still perhaps grant Czolgosz a posthumous scientific pardon.

He never got the chance. Spitzka removed the brain at 9:45 a.m., noting its warmth—the body can reach 130°F during electric execution. He sketched it while it cooled, then began to investigate every fold and fissure. As with Guiteau, the brain looked normal, unnerv-

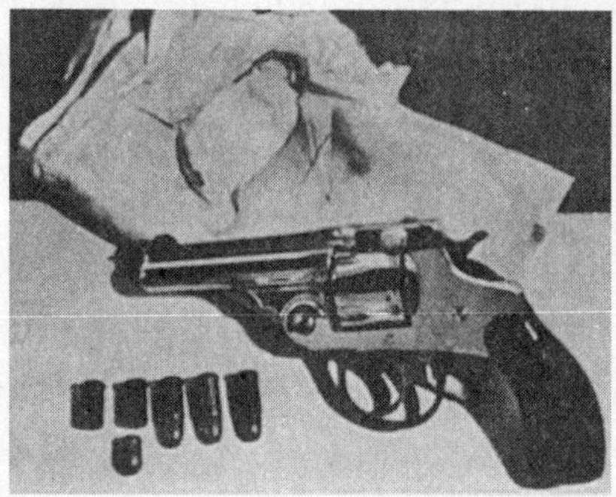

Left: Assassin Leon Czolgosz. Right: The gun and handkerchief Czolgosz used in the murder of McKinley.

ingly normal, on a gross scale. But before Spitzka could examine it microscopically, the prison warden stepped in. The warden had already received offers of $5,000 for Czolgosz's skull, and rather than risk making Czolgosz a martyr, he was determined to destroy every last scrap of him. Cruelly, spitefully, he refused Spitzka's plea for even a tiny slice of brain to examine later. Instead, the warden ordered the body sewn up at noon. He salted Czolgosz's corpse with barrels of quicklime, then poured in gallons of sulfuric acid. Based on experiments he'd been conducting with shanks of meat, he figured Czolgosz would liquefy in twelve hours. By midnight, Leon Czolgosz's troubled brain was no more.*

Like his father, then, the young Spitzka couldn't salvage the assassin's reputation. But neuroscience hadn't yet had its final say. As a good, sober scientist, the younger Spitzka admitted in the official autopsy that he'd found no sign of insanity. But in summing up, he added a qualifier: "some forms of psychoses," he wrote, "have no ascertainable anatomical basis... These psychoses depend rather upon circulatory and chemical disturbances."

Chemical disturbances. Even if the brain's anatomy appears normal, in other words, the brain still might not function properly, because of chemical imbalances. Spitzka's intuition here proved remarkable. To understand Guiteau's mental troubles, neuroscientists had to examine his cells. To understand Czolgosz's, they would have to drill down even deeper.

~

Smack in between the trials of these two American assassins, Santiago Ramón y Cajal had revealed that neurons were discrete cells. As a corollary, they must have a teeny gap, now called a synapse, separating them. But how exactly neurons sent signals across the gap—with pulses of chemicals or pings of electricity—remained unknown. Adherents of each idea called themselves "soups" and "sparks," respectively, and their mutual acrimony would shape the next half century of neuroscience.

The sparks had the upper hand at first. Electrical transmission seemed fresh and modern, chemical transmission passé, like those hoary old Greek theories about the four "humors." There was experimental evidence for electricity as well. Recently invented probes, fine enough to slip inside individual cells, revealed that neurons always discharged electricity whenever they fired. This was only an internal discharge, but it stood to reason that neurons would use electricity externally as well, to communicate with each other.

A set of macabre experiments with frog hearts seemed to buttress this theory further. By 1900 biologists knew that a frog heart, if removed from the frog and plopped into salt water, will beat *on its own* inside the solution. It just hovers there, throbbing, wholly disembodied yet somehow madly alive. Scientists discovered that they could even slow the heart rate down, or speed it up, by sparking different strands of the severed nerves that led into the heart. True, other scientists discovered that a splash of certain chemicals could accelerate or decelerate the heart in a similar way. But because the chemicals were man-made, the chemical action seemed a strange coincidence, little more.

A young scientist visiting England in 1903, Otto Loewi, found the frog-heart tricks fascinating, and upon returning to Austria he decided to investigate the link between nerves, electricity, and chemicals. Loewi, however, had an absentminded, daydreaming personality: when young he'd often blown off biology class to catch the opera or a philosophy lecture. So even though he became a noted pharmacologist, he neglected to follow up on the frog hearts. All the while the spark doctrine gained momentum.

Loewi finally got back into frog hearts in 1920, albeit under odd circumstances. The night before Easter that year, he nodded off while reading a novel. A Nobel-worthy experiment flashed before him in a dream, and he awoke, groggy, and jotted it down. The next morning he couldn't read his handwriting. Annoyed, then desperate, he pored over every jot and tittle. All he could remember was the moment of euphoria, the moment when everything made sense. He retired to bed crushed.

At three o'clock that night the dream returned. Loewi awoke and, rather than risk another loss in translation, scampered to his lab. There, he etherized two frogs, and slipped their cherry-sized hearts into two separate beakers of saline, where they beat and beat and made little waves against the glass. One heart had its nerves still attached, and when Loewi sparked certain nerve fibers, the beat slowed down, as expected. It was the next step that made him tingle. He sucked up saline from inside the first heart and squirted it into the other beaker. The second heart slowed down immediately. He then sparked some different nerve fibers on the first heart and sped it up. Another saline transplant made the second heart speed up, too—exactly as he'd dreamed. Loewi concluded that the nerve, whenever it was sparked, was spurting out some chemical. The chemical then got transferred to the second heart when he transferred the saline.

Loewi's experiment provided an enormous boost for the soups—proof that the nervous system, at least in some animals, did use chemicals to transmit messages. Other scientists quickly discovered heart-racing chemicals in mammals, then in human beings. After that the soup doctrine got so popular so quickly that Loewi collected a Nobel Prize for his dream work in 1936. (Typically blithe, though, he had to abandon the medal in 1938, leaving it behind in a bank vault: although he was Jewish, he'd paid no attention to the darkening thundercloud of Nazism, and when Hitler annexed Austria, he had to flee.*)

Still, Loewi and the soups had won only half the battle. The sparks conceded that the body might use chemical messengers in the peripheral nervous system, which controls mere limbs and viscera. But within the brain and spinal cord—the hallowed central nervous system—the sparks didn't budge. There, they insisted, the brain used electricity alone. And again, they did have good evidence for this, since neurons discharged electricity every time they fired. The sparks moreover argued that chemicals—the stuff of "spit, sweat, snot, and urine"—were too sluggish for brain duty. Only electricity seemed nimble enough, lightning enough, to underlie thought. Like Golgi's reticulists, sparks declared that the brain worked differently from the rest of the body.

But those who declare the brain somehow different, somehow biologically special, virtually always eat their words. Over the next few decades, the soups in fact detected plenty of chemicals that transmit signals *only* within the brain—so-called neurotransmitters. These discoveries undermined the hegemony of the sparks, and by the 1960s most scientists had integrated neurotransmitters into their understanding of how neurons work.

To wit: Whenever a neuron "fires," an electrical signal goes rippling down its axon to the axon tip—that's the electricity that the sparks had detected long ago. But electricity cannot jump between cells, not even across the 0.000001-inch-wide synapse separating one neuron from another. So the axon must translate the electrical message into chemicals, which can cross the chasm. Like a chemical supply depot, the axon tip stores and manufactures all sorts of different neurotransmitters. And depending on the message it needs to convey, the tip will package certain ones into tiny bubbles. These bubbles then dump their contents into the synapse, allowing the neurotransmitters to stream across the gap and connect with the dendrites of nearby neurons. This docking triggers those neurons to send an electrical signal down their own axons. At this point, with the message delivered, cleanup begins. Nearby glial cells start to remove excess neurotransmitter molecules from the synapse, either by vacuuming them up or releasing predator enzymes to shred them. This effectively resets the synapse so the neuron can fire again. All this happens within milliseconds.

Overall, you can think about the brain as both soup and spark, depending on what you measure and where—much the same way that photons of light are both waves and particles.

That said, the soup aspect has proved far more complex. The brain contains hundreds of types of neurons, all of which fire the same basic way electricity-wise. As a result, electrical signals can't convey much nuance. But neurons use over one hundred different neurotransmitters* to convey various subtleties of thought. Certain neurotransmitters (e.g., glutamate) excite other neurons, rile them up; other neurotransmitters (e.g., GABA) inhibit and anesthetize. Some brain processes even release both excitatory and inhibitory chemicals simultaneously. (When the brainstem sends us into dream sleep, for instance, it stirs up dreams by exciting certain neurons, but paralyzes our muscles by inhibiting others.) The neuron on the receiving end of a message must therefore sample the soup in a nearby synapse carefully, weighing every ingredient, before firing or not. The soup must taste just right to provoke the proper reaction.

~

The soup in Charles Guiteau's brain never tasted right. In retrospect, he almost certainly had schizophrenia, which disrupts neurotransmitters and skews their balance inside the brain—forcing neurons to fire when they shouldn't and preventing them from firing when they should. Syphilis also inflicted heavy damage. With his schizophrenia Guiteau was already on the brink, and when neurosyphilis started killing brain cells, his mind dribbled away into insanity.

Leon Czolgosz presents a tougher case. For one thing, it's nearly impossible to separate the judgments about his sanity from his era's dread of anarchy: some psychiatrists even defined anarchism itself as ipso facto mental illness. And while all five alienists who examined Czolgosz before his trial pronounced him sane, that sounds a little hollow when whole choruses of psychiatrists had sung the same line about Guiteau. Czolgosz's pretrial behavior doesn't clear up much, either. Czolgosz succumbed to a screaming fit one day in his cell, but some observers thought he'd shammed it. He once admitted that, after deciding to take McKinley down, "there was no escape" from the idea, not even "had my life been at stake"—but does that rise to the level of insane compulsion? And what about his habit of repeatedly wrapping a handkerchief around his hand in his cell? A guilty conscience? A mad tic? Depends on whom you ask.

Right after Czolgosz died, a few independent psychiatrists tracked down and interviewed family members and acquaintances, and they came away believing that Czolgosz had become unhinged not long before visiting Buffalo. One clue was that shooting the president seemed out of character. Czolgosz had no prior history of violence; in fact, patrons in bars often laughed at him for taking flies outdoors instead of swatting them. The psychiatrists noted, too, that Czolgosz barely understood anarchism and had converted to it only in May 1901—an awfully short time to become so fixated that you'd throw your life away, with no thought of escaping. And even fellow anarchists were baffled by Czolgosz's obsession with McKinley. The president had generally sided with management over labor during disputes, but he was no Rockefeller, no Carnegie, beating down the workingman, and McKinley himself never accumulated much lucre. (In one way, then, even Guiteau's ends seem more rational. Guiteau simply sought to install Chester Arthur in the White House. Czolgosz wanted to topple capitalism and the Republic in one swoop.)

Above all, the psychiatrists who studied the arc of Czolgosz's life emphasized how he'd changed after his mental breakdown and retreat to the family farm in 1898—becoming edgier, mistrustful, more isolated and paranoid. And it's here that Spitzka's comments about "chemical disturbances" seem most prescient. Czolgosz broke down in his midtwenties, a common time (as some historians have noted) for schizophrenia to emerge. I don't think that diagnosis holds up: Czolgosz was no Guiteau, untethered from reality. But given the primitive state of psychiatry in 1901, and the general rush to punish Czolgosz, the alienists might well have missed subtler symptoms, of subtler disorders, willfully or not. And regardless of the specific diagnosis, Czolgosz emerged from his breakdown a changed man: a desperately lonely man, someone longing for friends and meaningful work—but someone even anarchists, the most marginalized group in America, shunned. (In this he resembled less the hobnobbing, high-spirited Guiteau and more the loners Lee Harvey Oswald* and John Hinckley Jr.)

Sorting out cause and effect is tricky with brain chemistry: does depression cause changes in brain chemicals, or do changes in brain chemicals cause depression? The street probably runs both ways. But the balance of evidence does suggest that loneliness, isolation, and a sense of helplessness can all deplete neurotransmitters—can poison the soup and sap vital ingredients. That's surely part of what the younger Spitzka—after sluicing through a still-steaming brain on that cool October morning in 1901, hunting for signs of insanity and coming up empty—was getting at when he wrote about hidden chemical disturbances.

"I never had much luck at anything," Czolgosz once sighed, "and this preyed upon me." Indeed, it preyed upon him more than he knew: chronic stress can shrivel axons and dendrites, and skew the brain's thinking in unpredictable ways. That Spitzka intuited all this in 1901 is remarkable. And we can do even better today, since we know so much more about how neurons can affect global thinking patterns. We simply need to expand our scope and explore how individual neurons wire themselves together into circuits, which provide the raw material for our thoughts.

CHAPTER THREE

Wiring and Rewiring

We've seen how individual neurons work. But neurons often work best within larger and more sophisticated units called circuits—ensembles of neurons wired together for a common purpose.

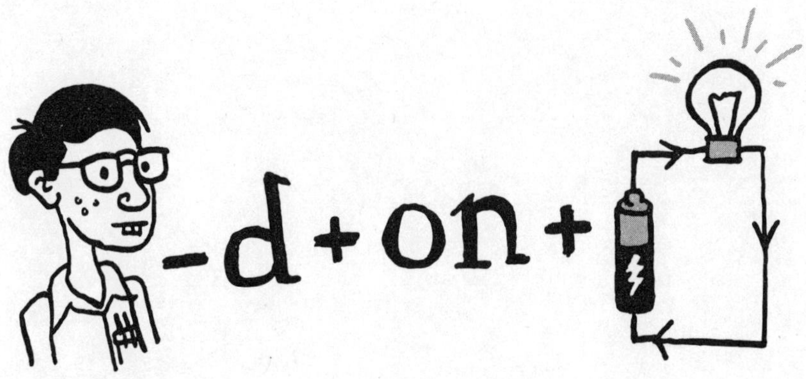

It was probably the most traveled outfit in history. A starched white shirt, a white cravat. Off-white button-up breeches. A deep-blue frock coat with brass buttons. An incongruous straw hat with a floppy brim. And most important, a metal-tipped hickory cane—the famous cane with which Lieutenant James Holman clicked his way through Siberia, Mongolia, Jerusalem, Mauritius, China, South Africa, Tasmania, Transylvania, and seemingly everywhere else in the known world.

Holman joined Britain's Royal Navy at age twelve, in 1798, and stayed active until just before the War of 1812, when he caught a mysterious illness off the coast of North America. Naval doctors, stumped by his roaming joint pain and headaches, diagnosed "flying gout," a meaningless catchall syndrome. However fictitious, flying gout handicapped Holman and forced him out of the navy at twenty-five.

While adjusting to his new, sedentary life in England, he earned an appointment as a Naval Knight of Windsor—which sounded grand, but in reality meant drear and tedium. His only duty was attending chapel twice a day and saying extra prayers for the king, his lords, and various toadies around Windsor Castle. The rest of the time he sat around his small apartment, alone, and did nothing; he couldn't even read. Holman found life at Windsor such an existential torture that his physical health deteriorated, and a case of wanderlust gripped him. He soon fled England and spent much of the rest of his life roaming, plunging headlong into odd and often dangerous corners of the globe.

For one early trip, he got it into his head to cross Siberia. Thanks to the atrocious, teeth-rattling ruts on the roads, he ended up walking much of the way—strolling alongside the cart, holding on to a rope. But before reaching the Pacific he got kidnapped by officials of the tsar and deported as a spy, since no one believed that anyone would travel Siberia for fun. On later trips he chased down slave traders;

mapped the Australian Outback; negotiated with headhunters; dodged forest fires; charged into war zones; and crossed the Indian Ocean on a ship carrying a cargo of sugar and champagne (it wasn't all hardship). He also climbed Mount Vesuvius in mideruption, a jaunt that nearly burned the soles off his shoes but proved to him that he could handle anything, despite his handicap. Along the way he earned a reputation as quite a ladies' man, and he did good enough scientific work (on the drifting of seeds among islands) to get elected to the Royal Society and cited by Charles Darwin. He rarely traveled in luxury—his pension amounted to just £84 yearly—and he stretched that further by packing his own meals (usually fruit, wine, and tongue, a cut-rate meat that didn't spoil) and by wearing his old naval uniform everywhere. Altogether Holman and his naval jacket, straw hat, and cane traveled 250,000 miles*—equivalent to ten trips around the equator or one trip to the moon, making him the most prolific traveler the world had ever known.

He returned to England as little as possible, and whenever he did find himself marooned at home, he took advantage of his downtime to write travel books. Eclectic and rambling, they might include soy sauce recipes on one page, advice for hunting kangaroos on the next, and he was constantly quoting the many poems he'd had to memorize. (He also included plenty of gossip about robberies, affairs, and local customs like sponge baths.) Before Holman even finished writing a book, though, that old desire to roam would well up inside him. Indeed, with his first book, published in 1822, he scurried to leave England almost before he'd finished the page proofs. The book became a bestseller, but by the time the London literati got hold of it and could flip to the frontispiece to see a portrait of this curious author, Holman was a thousand miles distant.

Holman couldn't have known it, but that frontispiece, while handsome overall, had one unsettling feature: his eyes, which seemed to look in different directions. Later portraits were even less flattering. In one book's frontispiece he looked drugged, with eyes unfocused. A later oil portrait showed him with an unsightly Rip Van Winkle beard and, again, vacant white eyes. In another depiction Holman is shown with his hand draped over a blank white globe, as if embracing a giant, irisless eyeball. Portraying him with a globe devoid of all features seems baffling at first, since Holman had covered more of the earth's surface than anyone alive. In truth, the blankness was fitting. Holman, you see, was blind.

His health troubles had started in the navy. His ship's patrol route yo-yoed endlessly between Nova Scotia, where the wind practically froze icicles inside men's noses, and the Caribbean, where the sunlight beat down hot enough to melt candles. Something about those extremes ruined his joints, and his ankles grew so stiff and sore that he could no longer pull on his boots, much less negotiate the rolling decks. He took shore leave and recovered, but more bitter nor'easters and more wilting afternoons finally broke him. Soon his eyes began to ache something awful: mere sunlight felt like needles piercing his retinas. His world gradually fell dark, and even though his doctors treated his eyes with leeches, poultices, opium, and lead ointments, nothing could rescue his vision. Segments of his optic nerves* finally died when he was twenty-five, severing that connection to the brain and leaving him permanently sightless. He would eventually set foot in nearly every country on earth, but would set eyes on none of them.

Holman almost never got the chance to travel, thanks to his pseudo-ennoblement. Naval Knight bylaws said that he and his six fellow knights could not be absent from England more than ten days per year. Holman obeyed at first, but the monotony of life at Windsor proved unendurable, and after just a few months there his fevers returned and his flying gout began clawing at him again. He needed activity, stimulation, and his doctors begged the two wardens of the Naval Knights to let him catch the next ship out. The wardens, sympathetic at first, let Holman go, and the travel worked wonders. When he returned to Windsor, though, and the tedium descended, his aches and pains started to torment him anew. He got another travel visa and immediately felt better. But the illness started right up again after the next homecoming. And the next, and the next. Writing books allayed the pain a little—memory is a powerful analgesic—but each time he finished a manuscript he felt worse, and needed more leave to recover. After Holman missed some state funerals and coronations, the Windsor wardens began to grumble.

They weren't the only ones. After each book appeared, pundits would challenge the very idea that a blind man had, or even could have, traveled so widely. As we'll see, modern neuroscience lends credence to Holman, but in the early 1800s, society treated blind people pretty shabbily. Most blind folk simply scrounged up a bowl and started begging for farthings. The luckier ones (arguably) worked in traveling carnivals, where they were dressed up in donkey ears and/or huge fake eyeglasses and shoved onstage. There, they stumbled around without any real script; the entertainment lay in watching the production fall to shambles. Beggars and buffoons were what people thought of when they thought of the blind, not circumnavigation and adventure.

Even those who didn't dismiss Holman condescended to him. "I am constantly asked..." he once wrote, "what is the use of travelling to one who cannot see?" Some idiots questioned whether Holman had really left England, since all seven continents must look the same to him. Holman gritted his teeth and explained that foreign lands sounded different, smelled different, had different weather patterns and different daily rhythms. And in fact, Holman rarely neglected other senses in his writing. Timbers squeal and crockery gets smashed and ships pitch about seasickeningly in storms. Holman eats monkeys "cooked in the manner of an Irish stew" and describes touching everything from snakeskin to statues in the Vatican museum. You don't need two good eyes to describe the horrors of dysentery, or of swarms of flies and mosquitoes so thick that he needed chain mail to protect himself. And in some ways, Holman argued, his handicap made him a *superior* traveler:* instead of relying on a superficial view of a scene, his blindness forced him to talk to people and ask questions.

Still, Holman did have some practical tricks, tactics for navigating a world he couldn't see. Instead of indistinguishable paper bills, he demanded coins for currency. He acquired a special pocket watch whose hands he could trace without interfering with its ticking. To record his observations, he used an inkless dictation machine called a Noctograph,* a wooden slab with wires strung every half inch to guide his hand across the paper. And in exchange for free passage on ships, he often bartered his services—especially storytelling, like Homer of old—to relieve the tedium of ocean travel. One story he no doubt related involved a short excursion (1,400 miles) that he took with a friend—who happened to be deaf. "The circumstance was somewhat droll," he later wrote. "We were not infrequently exposed to a jest on the subject, which we generally participated in, and sometimes contributed to improve." All travelers need a sense of humor.

Blind explorer James Holman. Notice the unfocused eyes and Noctograph dictation machine.

Perhaps most important, James Holman succeeded in traveling the world by himself because he took advantage of neuroscience. Like most blind people, Holman explored his immediate environment with his hands. (For this reason women found Holman alluring—they adored his heightened sense of touch and often granted him permission to "look over" their faces and even bodies.) For navigating the world at large, however—for dodging poles and trees, for negotiating crowded bazaars—Holman relied not on his hands but on his hickory cane. He didn't use the cane the way blind people do today, as a sort of extended finger to feel his way along. His cane was too short, too heavy, too inflexible for that. Instead, he clicked the metal nib onto the pavement every few steps, and listened.

Whenever he clicked the cane, sound waves ricocheted off any nearby objects, and the echoes arrived back in each ear at slightly different times. After some practice, his brain learned to triangulate those time differences and determine the layout of the scene confronting him. The echoes also revealed details about an object's size, shape, and texture—hard, skinny statues sound different from soft, broad horses. Mastering this sensory capacity—called echolocation, the same sense that bats use—took years of determined work, but determination was James Holman's long suit. And once he'd perfected it, he could navigate everything from Vatican art galleries to Mount Vesuvius mideruption. Like flicks of a flashlight in a dark room, these cane clicks became Holman's sight.

Scientists often call the human brain the most elaborate machine that ever existed. It contains some hundred billion neurons, and the tip of an average axon wires itself up to thousands of neighbors, producing an inordinate number of connections for analyzing data. (There are so many connections that neurons seem to obey the famed "six degrees of separation" law: no two neurons are separated by more than six steps.) And cases like James Holman's reveal even more intricacies, since they show how the human brain can deviate from the standard wiring plan and sometimes even rewire itself, by changing its wiring patterns over time. Some of these changes sound as fantastical as a blind man climbing volcanoes, but all of them give us insight into the incredible plasticity of our neural circuits.

~

To see how brain circuits work, imagine a noise—like a *clack* on a cobblestone—arriving at James Holman's ear. The *clack* vibrates various bones and membranes inside his ear canal, and the sound wave eventually transfers its energy to a fluid in his inner ear. That fluid sloshes over rows of tiny hair cells and (depending on the sound) bends some of them to a greater or lesser extent. These hairs are connected to the dendrites of nearby nerve cells, which immediately fire and transmit electrical signals down their long axon "wires" toward the brain. Upon reaching the brain, the signal causes the axon to squirt a chemical soup into a nearby synapse. This finally arouses neurons in the auditory cortex, a patch of gray matter in the temporal lobe that analyzes the sound's pitch, volume, and rhythm.

Reaching the auditory cortex is only the start, though. For Holman to consciously recognize the *clack* or navigate with it, the signal has to circulate to other patches of gray matter for further processing.

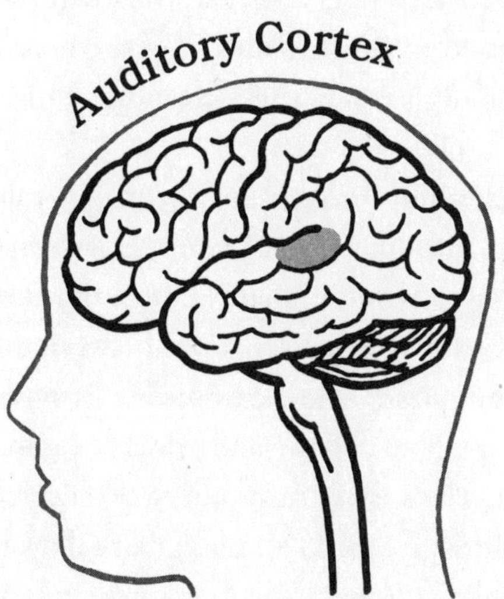

And reaching those other patches of gray matter requires going subterranean—diving beneath the gray matter surface and into the brain's white matter.

White matter consists largely of high-speed axon cables that zip information from one gray matter node to another, at speeds up to 250 miles per hour. These axons can shuttle information around so quickly because they're fatter than normal axons, and because they're sheathed in a fatty substance called myelin. Myelin acts like rubber insulation on wires and prevents the signal from petering out: in whales, giraffes, and other stretched creatures, a sheathed neuron can send a signal multiple yards with little loss of fidelity. (In contrast, diseases that fray myelin, like multiple sclerosis, destroy communication between different nodes in the brain.) In sum, you can think about the gray matter as a patchwork of chips that analyze different types of information, and about the white matter as cables that transmit information between those chips.

(And before we go further, I should point out that "gray" and "white" are misnomers. Gray matter looks pinkish-tan inside a living skull, while white matter, which makes up the bulk of the brain, looks pale pink. The white and gray colors appear only after you soak the brain in preservatives. Preservatives also harden the brain, which is normally tapioca-soft. This explains why the brain you might have dissected in biology class way back when didn't disintegrate between your fingers.)

A message traveling through a white-matter cable can either stir other neurons to life (pay attention!) or anesthetize them (pay no attention!). But given the inordinate number of neurons we have, and given the bazillions of pathways that run between different patches of neurons, one key question in neuroscience is how the *clack* signal "knows" which path to follow, and which neighbors to excite and which to inhibit. The answer turns out to be fairly simple: like James Holman's cart through Siberia, brain signals follow ruts.

Start with two neurons. If one neuron causes another to fire in quick succession over and over, the synapse between them actually changes in response. The axon tip of neuron one swells larger and starts packing in more bubbles of neurotransmitters to flood the synapse between them; wholly new axon branches might even sprout up. Neuron two can then make listening to neuron one a priority by extending more dendrite receptors back toward it. This allows neuron two to respond to even low-intensity prompts. Overall, just as a wagon wheel will carve a rut into the road after repeated journeys, repeated neuron firings will carve ruts into the brain that make signals much more likely to follow some neural tracks than others.

Scientists use a different metaphor to explain how neural connections grow stronger over time: *neurons that fire together wire together.* And usually it's not just two or three neurons firing and wiring together. Once a rut gets established, circuits of many thousands of neurons will fire in sequence.*

Thanks to white matter cables, these circuits can link together even distant patches of gray matter, allowing the brain to carry out complicated actions automatically. We're all born with circuits in our lower brain, for example, that control reflexes like sneezing, gagging, and yawning: as soon as the first neurons in the sequence fire, all the others follow, like a row of dominoes. That's why the steps involved in a sneeze or yawn rarely vary. Circuits in the higher brain work in the same way. After tons of practice, we all learned to link the letters d-o-g in our *Dick and Jane* primer with both an image of a fuzzy quadruped and with the sound *duh-aw-guh.* Eventually, any one of that triad automatically evokes the others. Negative experiences can wire neurons together, too. Enter an alley where you once had a fright, and its smells and shadows will reawaken your terror circuits.

All human brains subscribe to a standard wiring plan, which ensures that certain patches of neurons can always talk to certain other patches—and good thing. Your eyes better be able to rouse your fear circuits, and your fear circuits better be able to tell your legs to skedaddle, or you won't last long outdoors. This general wiring scheme gets laid down during our fetal days, when axons begin to bud and grow like shoots. That said, the general wiring diagram can vary in its details from person to person. One dramatic example of this is synesthesia, a condition in which people's senses blend together in trippy ways.

For most people, one sensory input produces just one sensory experience. Cherries simply taste like cherries, and rubbing sandpaper on the skin simply feels scratchy. For people with synesthesia, one sensory input leads to multiple outputs—the expected cherry taste, plus, say, a phantom tone. These superadded sensations are involuntary and consistent: each time the synesthete hears G-sharp, the exact same unaccountable pepper smell floods her nose. Synesthesia is idiosyncratic as well: while one person always sees the number 5 as fuchsia, another insists it's key-lime-pie green.

The most common type of synesthesia produces a symphony of color, especially when people hear certain sounds or see certain letters and numbers. Richard Feynman saw ecru *j*'s, indigo *n*'s, and chocolate *x*'s inside equations. Vladimir Nabokov once said that for him, the long vowel *aaah* has "the tint of weathered wood," while the shorter *ah* "evokes polished ebony." Franz Liszt used to berate his orchestra—who could only stare back, bewildered—for playing his music the wrong color: "Gentlemen, a little bluer, please, the tone depends on it!" Another time he implored: "That is a deep violet [passage]!... Not so rose."

Color-sound and color-letter synesthesia are the most common types because of brain geography: some of the regions that analyze sounds, letters, and colors lie right near each other, so signals can easily leak across the border. In theory, though, synesthesia can link any two sensations in the brain, and sixty known types exist. Hearing-motion synesthetes might hear a siren song emerging from a simple screen saver of moving dots. Touch-emotion synesthetes might feel silk as calming, oranges as shocking, wax as embarrassing, and denim as morose (so much for your favorite jeans). To touch-taste synesthetes, wrought-iron fences might taste salty, or certain kinds of meat "pointy." (One man pouted before a dinner party that the chicken he'd fricasseed came out too "spherical.") Sexual synesthetes might see colored shapes floating in front of them during coitus. Color-time synesthetes can experience days of the week, months of the year, or even stages of life as a patchwork of shades and hues. Imagine listening to Jacques's "seven ages of man" speech from *As You Like It,* and watching a rainbow envelop the stage.

Synesthesia probably has a genetic component, since it runs in families and pops up in most cultures. Importantly, too, neurologists have ruled out the idea that synesthetes are just talking metaphorical jive, the way the rest of us speak of "loud shirts" and "sharp cheddar." These people's brains actually work differently, as tests reveal. One experiment involved filling a piece of paper with a bunch of alarm-clock fives (5), but also scattering a few blocky twos (2) in there. Normal people find it nearly impossible to pick out the 2s without hunting one by one. To synesthetes, each 2 pops out in Technicolor, instantly. (It's similar to the way numbers pop out automatically on color-blind tests.) As another trick, if you show a synesthete, say, a giant numeral 4 made up of rows and rows of tiny 8s, the figure's color will flip depending on whether she focuses on the whole (the 4) or the pixels (the 8s). Other tests make synesthetes squirm. Normal people have no trouble reading text of basically any color. For synesthetes, numbers or letters that are the "wrong" color can disorient or repel them, since the colors on the page do battle with the colors in their minds.

```
88    88
88    88
88    88
888888888
      88
      88
      88
```

Neuroscientists know in a general way how synesthesia must work: the neuron circuits that process one sense must be accidentally strumming the circuits that process another sense, causing both sets to hum simultaneously. Determining exactly why that happens, though, has proved tricky. Two possible explanations have emerged, one anatomical, one functional. The anatomical theory blames poor pruning of neurons during childhood. All babies have far more neurons than they need; their neurons also have an excessive number of axon and dendrite branches. (As a result, young children probably experience synesthesia all the time.) As children develop, certain neurons begin firing together and wiring together, and those active neurons remain healthy. Meanwhile, unused neurons starve and die off. Excess branches get pruned back as well, like a maple near a power line. This destruction sounds brutal—neural Darwinism—but it leads to tighter, stronger, more efficient circuits among survivors. Perhaps the brains of synesthetes don't prune well. Perhaps their brains leave extra connections in place that link different sensory regions.

The functional theory suggests that neurons get pruned just fine, but that some neurons can't inhibit their neighbors very well. Again, our highly connected neurons have to discourage signals from shooting down stray paths to the wrong parts of the brain; they do so by skunking certain neighbors with inhibitory chemicals. But even if those stray paths lie dormant, they still exist—and could, in theory, open up and become active. Perhaps, then, the brains of synesthetes fail to inhibit these underground channels, and information leaks from one brain region to another.

The first clue for deciding between the functional and anatomical theories came from a Swiss chemist. In 1938 Albert Hofmann's drug company was searching for new stimulants, and he began investigating some chemicals derived from a fungus. He soon drifted to other compounds but had a nagging feeling that the fungi had more to teach him. So on a Friday afternoon in April 1943, he whipped up a fresh batch of one chemical, called lysergic acid diethylamide (in German, Lyserg-Säure-Diäthylamid). During the synthesis he suddenly felt woozy and saw streaks of color. He later guessed he'd gotten some powder on his finger, then rubbed his eyes. But he wasn't sure, so he tested his guess on Monday, April 19—forevermore known as Bicycle Day. He dissolved a tiny amount of powder, a quarter of a milligram, in a quarter shot of water. It had no taste, and down the hatch it went. This happened at 4:20 p.m., and although Hofmann tried to record his sensations in his lab journal, by five o'clock his handwriting had deteriorated into a scrawl. His last words were "desire to laugh." Feeling unsettled, he asked his assistant to escort him home on his bicycle. It was quite a trip.

On the ride, the streaks of color reappeared before his eyes, and everything became elongated and distorted, as if reflected in a curved mirror. Time slowed down as well: Hofmann thought the trip took ages, but the assistant remembered furious pedaling. In his drawing room at home, Hofmann struggled to form coherent sentences, but finally made it clear that (for some reason) he thought milk would cure him. A neighbor woman patiently hauled bottle after bottle to him, and he chugged two liters that night, to no avail. Worse, Hofmann began having supernatural visions. His mind transmogrified the neighbor into a witch, and he felt a demon rise up inside him and clutch his soul. Even his furniture seemed possessed, trembling with menace. He felt certain he'd die right there on his couch.

Only hours later did he calm down, and he actually enjoyed the last hour. His eyes became veritable kaleidoscopes, with *Fantasia*-like fountains of color "exploding [and] rearranging and hybridizing themselves in constant flux." It also pleased him, he later reported, that "every acoustic perception, such as the sound of a door handle or a passing automobile, became transformed into optical perceptions. Every sound generated a vividly changing image, with its own consistent form and color." In other words, the drug produced synesthesia, something he'd never experienced.

Hofmann's Lyserg-Säure-Diäthylamid eventually became known as LSD, and since then thousands of Phish and Grateful Dead fans have had similar experiences. Tripping on LSD obviously can't change the brain's hardwired circuits. LSD can interfere with neurotransmitters, however, and warp the information flowing through those circuits for a few hours. It's like flipping your television from a Ken Burns documentary to a David Lynch nightmare sequence—the same circuitry is providing the picture, but the content is much wilder. This provides strong support for the functional theory of synesthesia. There's some evidence that natural synesthetes still might have brains that are wired a little differently. But the experience of Hofmann and others suggests that we all might have a talent for synesthesia latent inside us, if only we could tap it.

~

Hofmann's drug-induced synesthesia showed that certain experiences can alter the flow of information through our neuronal wires, at least temporarily. But can any experiences actually *rewire* brain circuits in a permanent way?

Children's brains can remodel themselves quite easily and form all sorts of new connections: that's how they sponge up language and so much else. For most of the past century, though, neuroscientists considered remodeling in the adult brain impossible, thanks in part to Santiago Ramón y Cajal. Cajal spent a decade injuring the nerves and neurons of animals to test how well those tissues recovered. He found that peripheral nerves could often regenerate themselves (which explains why surgeons can reattach severed hands, feet, and penises, and get them working again). But neurons in the adult brain never grew back. This led Cajal to make the bleak declaration that "in the adult brain, nervous pathways are fixed and immutable. Everything may die, nothing may be regenerated."

Other observations supported Cajal's pessimism. Compared to children, adults have a much tougher time learning new skills like languages, a sign of neural sclerosis. And if adults suffer strokes or other brain damage, they might lose certain skills permanently, since neurons never grow back. Moreover, the lack of adult plasticity made sense from an evolutionary perspective. If the adult brain changed too easily, the thinking went, circuits controlling important behaviors and memories would unravel, and skills would evaporate from our minds. As one scientist observed, a fully plastic brain "learns everything and remembers nothing."

All that's true. But neuroscientists were a little hasty in declaring that the soft, pliable clay of the infant brain always gives way to sturdy but brittle ceramic. Even if the adult brain cannot grow new neurons* or repair damaged ones, that doesn't mean that all neuron *pathways* are fixed and immutable. With the right training, neurons can indeed change how they behave and transmit data. Old brain wires can learn astounding new tricks.

In the late 1960s, a degenerative eye disease claimed both retinas of a sixteen-year-old Wisconsinite named Roger Behm, rendering him blind. Forty years later he took a flier on a "vision substitution" device that a local scientist had built. The device consisted of a black-and-white video camera mounted on Behm's forehead, with a ribbon of wires leading down into his mouth. The wires ended in a rectangular green electrode, not much bigger than a postage stamp, that rested on Behm's tongue. The camera fed its images to this electrode, which transformed each pixel into a buzz of electricity reminiscent of seltzer bubbles: white pixels tingled his tongue a lot; black pixels gave no tingle; gray were intermediate. Behm was supposed to use the tongue "image" to interact with the world around him.

As you might expect, this flummoxed him at first. He nevertheless learned how to detect motion versus stillness rather quickly. He started picking out triangles, circles, and other Euclidians not long afterward. He graduated to common objects like cups, chairs, and telephones. Soon he could pick out logos on football helmets and sort playing cards by their suits, even navigate a simple obstacle course. Nor was Behm unique or special in picking up these skills. Other blind people learned how to use mirrors, pick out overlapping objects, or follow the writhing dance of a candle flame.

The man behind the device, Paul Bach-y-Rita, became a neuroscientist in a roundabout way. (Although a Bronx native, Bach-y-Rita had a compound Catalan surname, like Santiago Ramón y Cajal.) Bach-y-Rita attended medical school in Mexico City on a dare, then dropped out to work, among other itinerant jobs, as a masseur and a fisherman in Florida. He also taught anatomy to blind people who were studying to become masseurs, which helped him understand how they interacted with the world. (The blind, with their heightened sense of touch, make fine masseurs and masseuses.) Eventually he returned to medical school and started working with blind patients. But Bach-y-Rita really found his purpose in life after his father, Pedro, suffered a massive stroke in 1959 and was left half paralyzed and speechless.

Pedro entered a rehabilitation clinic, but when his progress plateaued, his doctors declared him doomed and suggested a nursing home, since his fixed and immutable brain would never recover. This fatalism—so common in rehab facilities then—angered Bach-y-Rita's brother, a doctor named George. So George designed his own rehab regimen. It sounded harsh: George made Pedro crawl like an infant at first, learning how to move each limb again, before gradually working him up to his feet. He then made Pedro do household chores such as sweeping the porch and scrubbing pots and pans. Pedro struggled mightily and appeared to make little progress, but the repetitive motions eventually retrained his brain: he not only regained the ability to talk and walk, he resumed his teaching job, remarried, and started hiking again. Pedro in fact died (seven years later, of a heart attack) while hiking in the mountains of Colombia, at age seventy-three. His autopsy revealed extensive lingering damage, especially to white matter cables that connect certain patches of gray matter to each other. Importantly, though, the gray matter itself still worked. And his brain proved plastic enough to reroute the cues for walking and talking around the ravaged tissue. That is, instead of routing signals from A to B, it now routed them from A to C and then C to B—not the most efficient path, but one that improved over time as the mental ruts grew deeper.

Inspired, Paul Bach-y-Rita did additional residencies in neurology and rehab medicine, and decided to investigate brain plasticity himself, especially how blind people might regain a vestige of sight. His first "brain port" used a hand-cranked video camera; it projected an image onto the viewer's back via vibrating Teflon studs implanted in a dentist's chair. With just four hundred pixels, the images looked like a black-and-white television with poor focus. Nevertheless, with practice people could pick out individuals based on their hairstyles and faces, including 1960s supermodel Twiggy. (The patients shrugged when shown *Playboy* centerfolds, however—touch still beats sight in some areas.)

When microprocessors got small enough, Bach-y-Rita built devices to stimulate the tongue, one of the body's most sensitive tactile areas. (Saliva also makes the mouth more conductive than bare skin, lowering the necessary voltage.) And the devices really gained legitimacy when scientists started scanning the brains of patients while they used them. The scans revealed that, even though the video information came streaming in through the tongue, the brain's vision centers crackled with activity. Neurologically, this input was indistinguishable from "sight." Psychologically, too, the patient experienced the tactile tongue data as vision. Blind people using the devices perceived objects as being "out there" in space in front of them, not on their tongues. They flinched from balls flung at them, and could sense when objects moved closer or farther away because they grew larger or smaller. They even fell prey to certain optical illusions, like the "waterfall effect." If you stare at something in motion (like a waterfall) for several seconds and then look away, whatever you focus on next seems to move of its own accord. Bach-y-Rita's device induces this same vertiginous feeling in blind people, further proof of a latent neurological ability to see.

Meanwhile, Bach-y-Rita's team developed other sensory substitution devices. A leper who'd lost the sense of touch in his hands (leprosy destroys nerves) donned a special glove that piped tactile information to his forehead instead; within minutes he could feel the cracks on a table and distinguish between rough logs, smooth aluminum tubes, and soft rolls of toilet paper. Bach-y-Rita also worked on "electric condoms." Many paralyzed men can still get erections, even if they can't feel them, and Bach-y-Rita's device, if ever completed, would pipe electric orgasms into their brains.

Most dramatically, Bach-y-Rita's team has restored people's sense of balance. This work started with a thirty-nine-year-old Wisconsin woman named Cheryl Schiltz, who'd taken an antibiotic called gentamicin after a hysterectomy in 1997. Gentamicin fights infections well but has a nasty habit of destroying the tiny hairs in the inner ear that keep us balanced and upright. Although these hairs are located in different tubes than the hairs that help us hear, they work the same basic way. A gel inside the tubes sloshes back and forth like jiggled Jell-O as our heads tip this way and that. This causes the hairs embedded in the gel to bend to and fro and thereby trigger certain neurons. From this data the brain determines whether we're standing upright and then corrects for deviations. With those hairs destroyed, the balance center in Schiltz's brain (the vestibular nuclei) went on the fritz and started shooting out signals at random to her muscles, forcing her to sway side to side, with little jerks. Worse, she always felt on the verge of toppling over, even while she was lying down, like a permanent case of the drunken spinnies. Schiltz and other gentamicin victims call themselves Wobblers. Most can barely navigate their own homes much less brave the outside world, where a simple zigzag on a carpet can send them reeling. Not a few Wobblers commit suicide.

Although skeptical, Schiltz let Bach-y-Rita's team rig her up in a green construction helmet with a tiny balance and some electronics mounted inside. Like Behm's device, wires snaked down from the headpiece to an electrode in Schiltz's mouth. When standing tall and true, she felt a kazoo buzz on the center of her tongue. When her head drooped or swayed, she felt the buzz slide forward, backward, or sideways. Her goal was to shift her posture to keep the buzz in the center at all times. The buzz felt bizarre to her, but she got the hang of it quickly. After sessions of just five minutes, she found she could stand on her own for a few precious seconds. One day she drilled for twenty straight minutes and found she could walk without staggering. Further practice improved her balance still more, and eventually Schiltz dispensed with the helmet altogether. She even learned to jump rope and ride a bike again.

More poignantly, she began training others on how to use the device, including Bach-y-Rita himself. After being diagnosed with cancer in 2004, Bach-y-Rita took a chemotherapy drug that damaged his own inner ear hairs and wiped out his sense of balance. So Schiltz walked him through how to use the green helmet—returning the favor to him, and ensuring that he would walk on his own right up until his death in 2006.

Scientists are still debating exactly how sensory substitution devices changed the brains of people like Behm and Schiltz. One good guess is that these devices, in rerouting information from the tongue to the vision and balance centers, take advantage of pathways and feedback loops that already exist. When eating an apple, for instance, your brain naturally combines information about its taste, crunch, and shiny red finish to give you a more comprehensive understanding. So we already mix some sensory input, and maybe the tongue data getting transformed into visual data is just an extreme example. In addition, as LSD synesthesia shows, there are plenty of dormant, underground, pseudo-synesthesic channels inside the brain to exploit as well.

It seems that our brains, being partly plastic, can swap one sense for another no matter how it gets piped in. This has profound implications for how we understand the senses in general. From this point of view, all the ears, eyes, and nostrils really do is tickle certain nerves. As a result, all sensory input looks pretty much the same after it leaves the sense organ and enters the nervous system: it's nothing but chemical and electrical blips. It's really our neuron circuits, not our sensory equipment, that decipher incoming signals and conjure up perceptions.

Scientists have by no means resolved all the scientific issues here, much less the philosophical conundrums. And frankly, the debates surrounding these devices can get pretty Jesuitical—*can blind people ever truly see?* But according to Bach-y-Rita, "We don't see with the eyes, we don't hear with the ears. All of that goes on in the brain." If that's true, then blind people really can learn to see, whether through their tongues, like Behm, or through their ears, like James Holman and his latter-day descendants.

~

Bach-y-Rita exploited modern electronics to remodel people's brains, but in truth, we don't need anything so sophisticated to take advantage of neural plasticity. Echolocators can transform their brains with nothing more advanced than their own teeth, tongues, and lips.

The most famous living echolocator, Daniel Kish, lost both his eyes at thirteen months to retinoblastoma, eye cancer; his sockets are hollow scars. But at age two, all on his own, he discovered the power of echoes. He developed a way of click-click-clicking his tongue—like a gas stove, albeit slower—to shoot out exploratory sound waves. He now navigates by listening for the echoes that reverberate around him.

To see how this works, imagine Kish approaching an object while walking down the sidewalk. *Click-click-click.* He notices that his tongue clicks echo back from points near the ground, but that the echoes stop at about bellybutton height. A few steps on, the echoes bounce back up to chest height; a few more steps, and they drop again. That echo profile indicates a parked sedan. Similarly, telephone poles produce a tall and skinny profile. Sound quality also provides clues: whereas cars reflect noise sharply, bushes muffle it.

Kish can echolocate with enough agility to climb trees, dance, and ride his bike in heavy traffic. He also bought a twelve-by-twelve-foot cabin in Angeles National Forest, near his home, and then spent days alone there navigating trails and crossing streams on slippery rocks. Kish's abandon has gotten him injured at times—smashed teeth, a broken heel. He also woke up in his cabin one night to find it on fire (bad chimney) and barely escaped. But he calls these frights "the price of freedom." As he has written, "Running into a pole is a drag, but never being allowed to run into a pole is a disaster."* It's a sentiment James Holman would have hear-heared.

As a matter of fact, the feats of modern echolocators like Kish lend credence to Holman's life story. On brain scans, echolocators show strong activity in the visual cortex while they're listening to clicks. That's probably because vision neurons, in helping us see things, also help us navigate the world around us. So they'd naturally be recruited for echolocation even if the raw input is auditory. After years of listening to the echoes from his cane, James Holman's brain almost certainly remodeled itself in the same way. His auditory neurons and vision neurons had fired together so frequently and wired themselves together so intimately that translating sound maps into spatial maps became instinctual.

Unfortunately, Holman had fewer and fewer opportunities to exercise that instinct over the years. His health depended on travel, but as he began to request more and more leave time from the Naval Knights and began to travel farther and farther afield, and especially as he began to profit on his travels by publishing books—books full of exploits, such as climbing Mount Vesuvius, that seemed possible only for an able-bodied man—the wardens started seething. In retrospect, Holman probably had a psychosomatic illness: the depression that plagued his mind during his idle time in England also afflicted his body; conversely, traveling buoyed his spirits and relieved his physical pain. But with every trip the wardens of Windsor became more convinced that Holman was scamming them, and they began to forbid his travel, essentially sentencing him to house arrest. During these spells Holman appealed for help to every medical and political authority he could—a young Queen Victoria even got involved. But like Pharaoh of old, the wardens hardened their hearts and would not listen.

By 1855 Holman, in his midsixties, could barely manage a holiday to France anymore. And in truth, poor health was just one of several painful realities he had to face. When abroad, he kept wearing that staple of his traveling days, his naval uniform. But the coat and breeches had gone so out of fashion that even other sailors barely recognized him as a former officer. Worse, Holman's celebrity with the general public had dimmed. He published his last travel book in 1832, and year by year he fell deeper into obscurity. On the rare occasions when a contemporary did mention him, it was usually in the past tense.

After turning seventy Holman stopped traveling altogether and rarely left his apartment. Friends worried about him, but it emerged that he'd actually thrown himself into one last journey, into his past, to write his autobiography. The long hours he spent straining on the Noctograph dictation machine depleted him further, but he pushed himself because he imagined the book would secure his legacy at last. He still wanted recognition that his journeys had meant something—and meant something beyond the fact that a blind man had undertaken them. He saw himself not as a sightless Marco Polo but as Marco Polo's equal.

He completed the autobiography just before dying, in 1857. Sadly, no publishing house would take it, citing poor sales of his previous work. He left it to a literary executor, but that man in turn soon died, and within a few decades the book was lost to history.

Almost everything we know about Holman's personal life, then, comes from his surviving books—and it's not much. His favorite memories, his greatest disappointments, the names of his lovers, all of that remains unknown. He never even revealed how he first learned to echolocate. In fact, his travelogues spend amazingly little time discussing his blindness. Only one passage stands out for its frank discussion of his handicap and how it changed his worldview. In it, Holman was reminiscing about a few rendezvous from his past. Disarmingly, he admitted that he had no idea what his paramours looked like, or even whether they were homely. Moreover, he didn't care: by abandoning the standards of the sighted world, he argued, he could tap into a more divine and more authentic beauty. Hearing a woman's voice and feeling her caresses—and then filling in what was missing with his own fancy—gave him more pleasure than the mere sight of a woman ever had, he said, a pleasure beyond reality. "Are there any who imagine," Holman asked, "that my loss of eyesight must necessarily deny me the enjoyment of such contemplations? How much more do I pity the mental darkness which could give rise to such an error."

Holman was talking about love here; but in talking about desires and contemplations above and beyond what his eyes could strictly see, he was getting at something bigger—something bigger about himself, and about how all human beings perceive the world. With regard to sensory substitution, Paul Bach-y-Rita said, "We don't see with the eyes. We see with the brain." That sentiment is true in a broader sense as well. We all construct our reality to some degree, and if Holman augmented the scenes around him with his own imaginings, well, so do the rest of us. In other words, our neurons do more than simply record the world around us. As we'll see next chapter, neuron circuits actually wire themselves together into still larger units, allowing our brains to reinterpret and remake what we see—infusing simple sights with layers of meaning and coloring mere perceptions with our own desires.

CHAPTER FOUR

Facing Brain Damage

Circuits of neurons in turn combine to form larger structures, like our sensory systems, which analyze information in advanced ways.

A man lies along a table, a mask of plaster covering his face. The mask looks normal—nose, ears, eyes, teeth, lips. But when it's lifted, part of the soldier's face beneath seems to lift off with it, leaving a crater in his flesh. Sitting up, the soldier gets his first deep breath since the plaster was painted on a half hour before. Assuming he has a nose, he might catch the scent of the flowers kept nearby to brighten the Paris studio. Assuming he has ears, he might hear the clatter of dominoes across the room, from other mutilated soldiers waiting for their turn on the table. Assuming he has a tongue, he might sip some *vin blanc* to revive himself. And assuming he has eyes, he might see dozens of other masks hanging on the wall—the befores and afters of fellow-*mutilés* who'd lost their faces in the Great War and hoped the masks would help them resume a normal life.

The woman making the masks had no medical qualifications, only artistic ones. Although American, Anna Coleman Ladd had lived most of her early life in Paris and studied sculpture there in the late 1800s; Auguste Rodin himself had advised her. Still, she lacked the élan to be famous. She ended up carving staid satyrs and nymphs for fountains and private gardens, and she all but gave up sculpture when she returned to Boston to marry a Harvard medical professor. They had an unconventionally independent marriage, but Ladd followed him to Europe in 1917 and later snuck into Paris. Inspired by a similar outfit in London, called the Tin Noses Shop, Ladd opened her prosthetic mask studio in 1918 in a fifth-floor walk-up in an ivied building. She populated the courtyard below with her old busts and sculptures—sculptures with classically beautiful faces that, however passé to the art world, must have stirred the hopes of the *mutilés* sneaking in for appointments beneath the cover of dawn or dusk.

In one sense Ladd's studio was conducting an artistic experiment in the tradition of Pygmalion—how realistic could realism get? At the same time she was conducting a psychological experiment—could she fool the brain into mistaking a mask for flesh? We humans often conflate our faces with our very selves. So by restoring a face torn in two by a bullet, she was attempting to restore a soldier's identity. But she had no way of knowing whether other people—or the soldiers themselves—would accept the new faces as authentic.

Rebuilding a face wasn't something doctors often worried about before 1914. A few soldiers and brawlers in history—most notably the emperor Justinian II and astronomer Tycho Brahe—had lost their noses in sword duels. Most received silver or copper replacements, and some surgeons did develop "natural" methods for replacing lost tissue. (One involved sewing the face to the crook of the elbow for a few weeks, until the arm skin adhered to the bridge of the nose and provided a cover flap.*) But the trench warfare of World War I produced orders of magnitude more facial casualties than ever before, the result of grenades, mortars, machine guns, and other methods of flinging metal at high velocities. Just before going down, many soldiers heard a crackle or whistle from a shell, then felt their facial bones explode. One man compared the feeling to "a glass bottle [dropped] into a porcelain bathtub." Even dense jawbones might pulverize on contact, reduced to sand beneath the skin. And while metal helmets protected the brain, the helmet itself sometimes exploded into shrapnel when struck, gouging into eyes and ears. In all, tens of thousands of men (and a few women) woke up in a mudhole to find their noses torn off or tongues dangling. Some who lost eyelids slowly went blind as their corneas dried out. Other soldiers' faces looked punched-in, like a Francis Bacon portrait.* Officers instructed men on watch duty that, when peeking at the enemy, they should put their heads *and* shoulders over the parapet, since snipers would aim for the body, a more agreeable place to be shot.

The apocalyptic Battle of the Somme in 1916—when newspapers had to print not just columns but whole pages of casualties—spurred the British military to open a hospital for facial injuries on a dairy farm in Kent. The head surgeon there, a part-time painter, had seen how slapdash plastic surgery could be: he'd once encountered a young man in a POW camp who had hair growing on his nose because someone had grafted skin from his scalp onto his face. Determined to end such practices, the surgeon emphasized the aesthetics of facial reconstruction, even demanding multiple surgeries to get things right. In all, the Kent hospital performed eleven thousand surgeries on five thousand British soldiers, and often tended them for months between operations. Some victims could swallow only liquids, so the farm also raised chickens and cows and fed the men an eggnog slurry for protein. As part of their rehabilitation, some soldiers tended the animals, while others learned trades such as making toys, repairing clocks, or hairdressing. Many of the men formed deep friendships with their fellow "gargoyles," while others, being soldiers, also flirted with whatever women happened by. The boldest patients won their nurses as wives, and one exhilarated female visitor declared that "men without noses are very beautiful—like antique marbles."

Not everyone was so broad-minded. The soldiers felt safe enough while in the wards to tease each other, even call each other ugly; but they always wore red ties and cornflower-blue jackets when visiting the nearest village, to warn people off from a distance. Shopkeepers wouldn't sell the men liquor because some became unhinged when drunk, and outsiders dreaded eating with them because their food sometimes reappeared through extra holes when they chewed and swallowed. Some hospitals forbade the men mirrors, and when released from the safe cocoon of the facial ward, many patients killed themselves. Others found work in a new industry, enjoying long hours of dark solitude as cinema projectionists. And some of the direst cases, those that surgeons couldn't salvage, sought out Ladd or her London counterpart.

To sculpt a face, Ladd used a man's siblings or a preinjury photograph as models. A few hopeful soldiers brought in pictures of Rupert Brooke, a disarmingly handsome celebrity poet. Most, though, didn't care about being gorgeous. They wanted only to become anonymous again. As a first step, Ladd plugged any holes in their faces with cotton and painted plaster onto whatever portion needed masking. She sculpted the new features with clay, then created the actual mask a few days later by electroplating thin layers of copper and silver onto the clay surface. She might affix some absorbent pads behind the façade if the man's tear ducts or salivary glands leaked, but otherwise the six-ounce metal mask rested directly on the face, anchored by spectacles. She colored the masks with cream bath enamels to match skin tones, then made mustaches of metal foil, since real hair didn't adhere. Each mask took a month to produce, cost around $18 ($250 today), and could be cleaned with potato juice. Ladd's studio produced especially brilliant work. She painted gorgeous eyes, and left the slightest hint of blue in the cheeks to make them look freshly shaved. She also made foil mustaches so realistic that Frenchmen could twirl them (they much appreciated this), and even left their metal lips agape for cigarettes (ditto).

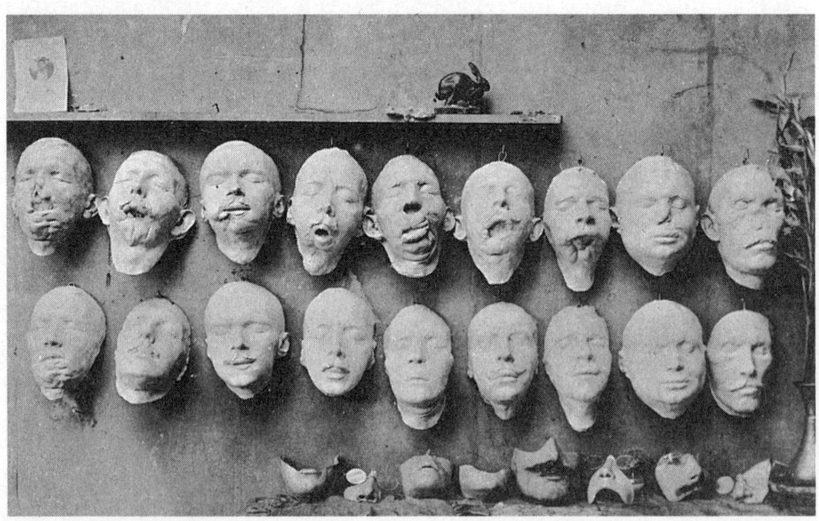

Plaster casts of soldiers' faces. Notice the finished, wearable masks on the bottom. (Library of Congress)

Ladd and her assistants made hundreds of soldiers achingly happy. "The woman I love no longer finds me repulsive," one lad wrote, "as she had a right to do." One veteran wore his mask during his wedding, and many more were buried in theirs over the next few decades. But however grateful, many found the masks too uncomfortable for daily use. The face has inordinate numbers of nerve endings, and the masks sometimes rubbed scars raw. Worse, the masks didn't function like real faces—didn't chew, didn't smile, didn't kiss. Even visually, the masks sometimes didn't cut it. The features didn't age the way skin did. The enamel chipped or corroded. And electric lighting, increasingly popular, often exposed the *Phantom of the Opera*–like seam between façade and flesh.

In the end, then, Ladd fell short: however artistic, her masks couldn't quite simulate the experience of seeing a real human face. As a result, the deeper, more psychological questions that her work raised—can the brain adjust to seeing a new face in the mirror? would that change someone's sense of self?—remained unanswered. It would take another century of work to get at those questions. And answering them would require understanding not only how the brain analyzes faces but, even more fundamentally, understanding how the brain sees the world around us at all.

~

The twentieth century's first major discovery about vision came about, once again, because of war. Russia had long coveted a warm-water port on the Pacific Ocean, so in 1904 the czar sent hundreds of thousands of troops to Manchuria and Korea to bully one away from the Japanese. These soldiers were armed with high-speed rifles whose tiny, quarter-inch bullets rocketed from the muzzle at fourteen hundred miles per hour. Fast enough to penetrate the skull but small enough to avoid messy shattering, these bullets made clean, precise wounds like worm tracks through an apple. Japanese soldiers who were shot through the back of the brain—through the vision centers, in the occipital lobe—often woke up to find themselves with tiny blind spots, as if they were wearing glasses spattered with black paint.

Tatsuji Inouye, a Japanese ophthalmologist, had the uncomfortable job of calculating how much of a pension these speckled-blind soldiers should receive, based on the percentage of vision lost. Inouye could have gotten away with merely showing them a few pictures and jotting down what details they could and couldn't see. But he was that rare thing, an idealistic bureaucrat, and he saw that his work revealed something deeper.

By 1904 neuroscientists knew a little about how vision worked in the brain. They knew that everything to the left of your nose (called the left visual field) gets transmitted into the brain's right hemisphere, and that everything to the right of the nose (the right visual field) gets transmitted into the left hemisphere.* Moreover, scientists knew that the occipital lobe was somehow involved with vision, since strokes back there often blinded people. But strokes caused such messy, widespread damage that the inner workings of the lobe remained mysterious. Scorching Russian bullets, in contrast, produced focal lesions as they entered and exited the brain. Inouye realized that if he could chart each man's specific brain damage, and match that damage to the part of the eye where a blind spot now appeared, he could basically produce a map of the occipital lobe—and thereby determine what sections of the brain analyzed each part of the visual field.

Before he got too far along with this work, Inouye stopped to examine a big assumption—that bullets followed a straight line through the brain. Perhaps they ricocheted around inside the skull, or got gummed up and followed a twisted path instead. So Inouye hunted down soldiers who'd been shot through the top of the head while lying on their stomachs. In this position the bullets ran parallel to their spinal cords. So in addition to a skull entry wound and a skull exit wound, most men also had, crucially, a third wound where the bullet left the skull and plugged them in the chest or shoulder. Inouye had the men re-create their postures the moment they'd been shot, and he found that all three wounds always described a straight line. Confident now that he hadn't overlooked anything, Inouye began mapping the occipital lobe, especially what's now called the primary visual cortex (PVC).

His most important finding was that our brains effectively magnify whatever we're looking at, by dedicating more neurons to the center of the visual field. Part of the primary visual cortex lies on the surface of the brain, just below the bump on the back of the noggin, and part of it lies buried beneath the brain's surface. It turns out that soldiers with black speckles in the center of their vision always had damage to surface patches, while men with peripheral speckles had damage to the subterranean stuff. The consistency of this correlation proved, as Inouye had hoped, that certain regions of the brain always controlled certain parts of the eye.

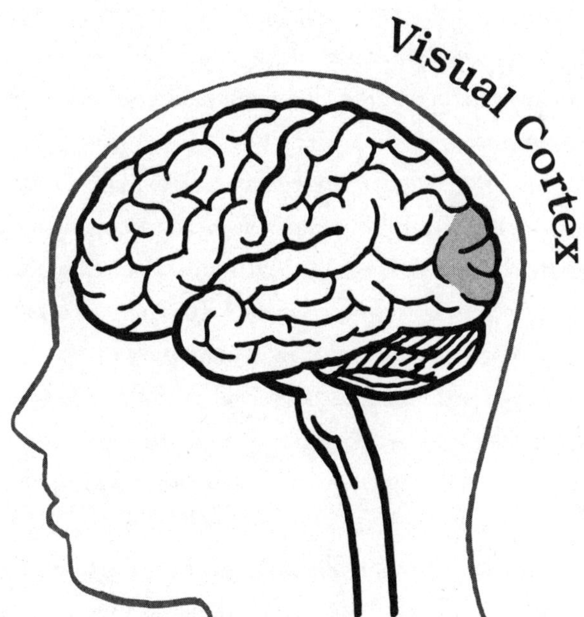

But the patches that processed the center, he discovered, were vastly larger in area than the patches that covered the periphery. In fact, it wasn't even close. Scientists now know that the focal center of the eye, the fovea, takes up just one ten-thousandth of the retina's surface area. Yet gobbles up a full one-tenth of the PVC's processing power. Similarly, around half of the PVC's 250 million neurons help us process the central 1 percent of our visual field. Inouye's half-blind patients helped him see this special magnification for the first time in history.

Unfortunately for Inouye, other scientists got credit for his discoveries. During World War I, two English doctors, ignorant of his work, duplicated his experiments on the visual cortex with their own brain-damaged troops. They obtained the same results, but these doctors had the cultural advantage of being European. What's more, in his major paper on vision, Inouye used a convoluted Cartesian graph to plot the relationship between the eyes and the primary visual cortex. It was precise, but it left readers cross-eyed themselves. The Englishmen meanwhile used a simple map, something scientists could grasp at a glance. When this intuitive diagram was published in textbooks worldwide, Inouye slipped into obscurity. Blindness can be a generational affliction, too.

The next major discovery in vision neuroscience took place far from the battlefield. In 1958 a pair of young neuroscientists at the Johns Hopkins University, one Canadian and one Swedish, began investigating neurons in the visual cortex. In particular, David Hubel and Torsten Wiesel wanted to know what sights or shapes got these neurons excited—what made them fire? They had a good hunch, based on other scientists' work. Signals from the eyes actually make a quick layover in the thalamus, in the center of the brain, before reaching the visual cortex. And other scientists had shown that thalamic neurons respond strongly to black-and-white spots. So Hubel and Wiesel decided to take the obvious next step and investigate how neurons in the visual cortex responded to spots.

When shown their new lab, a grimy basement with no windows, Hubel and Wiesel rejoiced. No windows meant no stray light—

perfect for vision work. They were less enthusiastic about the equipment they inherited. Their experiments involved, à la *A Clockwork Orange,* strapping down an anesthetized cat into a harness, immobilizing its eyes, and forcing it to stare at spots projected onto a bedsheet. But because the harness they inherited was horizontal, the kitty had to lie on its back, staring straight up toward the ceiling. Therefore the duo had to flip their slide projector toward the ceiling, too, then drape a sheet over the pipes up there "like a circus tent," Hubel remembered. Insects and dust rained down, and to see the screen, the duo had to stare upward themselves, straining their necks.

And this was just the setup—actually studying the neurons proved no easier. By 1958 scientists had built microelectrodes sensitive enough to monitor a single neuron inside the brain; some researchers had already examined hundreds of individual cells this way. (This head start intimidated Hubel and Wiesel, who felt like amateurs. So they "catapulted [themselves] to respectability," as they said, by starting the count in their experiments at number 3,000. Whenever people visited the lab, they made sure to announce what number they were on.)

Each electrode had fine platinum wires that slid into the cat's primary visual cortex. Hubel and Wiesel wired the electrode's other end to a speaker, which clicked whenever a neuron fired in response to a spot. Or at least it should have clicked. The first experiments proved dreadful, taking nine hours each—their necks were killing them—and running into the wee hours. Wiesel would start blathering in Swedish around 3 a.m., and Hubel almost nodded off and crashed one night while driving home. Worse, the neurons they monitored would not fire. They tried white spots. They tried black spots. They tried polka dots. "We tried everything short of standing on our heads," Hubel recalled—including cheesecake shots from glamour magazines. But the stubborn stupid neurons refused to click.

Week after maddening week passed, until September 1958. During the fifth hour of work one night, starting with cell 3009, they dropped yet another slide with yet another dot into the projector. According to different accounts, the slide either jammed or went in crooked, at an angle. Regardless, something finally happened: one neuron "went off like a machine gun," Hubel said—*rat-a-tat-tat-tat-tat-tat.* It soon fell silent again, but after an hour of desperate fiddling, they realized what was going on. The neuron didn't give a fig about the dot; it was firing in response to the slide itself—specifically, to the sharp shadow that formed on the screen as the edge of the slide dropped into place. This neuron dug lines.

More hours of fiddling followed, and the duo quickly realized how lucky they'd been. Only lines within about ten degrees of one orientation set this neuron off. Had they dropped the slide in any less crookedly, the cell would have continued to give them the silent treatment. What's more, other neurons, in follow-up experiments, proved equally picky, firing only for lines like \ or /. It took many more years of work, and many more cats, to firm everything up, but Hubel and Wiesel had already gotten a peek at the first law of vision: neurons in the primary visual cortex like lines, but different neurons like different lines, raked at different angles.

The next step involved looking a little wider and determining the geographical patterns of these line-loving neurons. Did all of the neurons that liked a given angle cluster together, or was their distribution random? The former, it turned out. Again, neuroscientists knew by about 1900 that neurons are arranged in columns, like bits of stubble on the brain's surface. And Hubel and Wiesel found that all the neurons within one column had similar taste: they all preferred one specific line orientation, like \. What's more, if Hubel and Wiesel shifted their platinum wire a smidge, about two-thousandths of an inch, to another column, all of that column's cells might respond to |, a line ten or so degrees different. Successive, tiny steps into new "orientation columns" revealed neurons that fired only for /, then ⁄ and so on. In sum, the optimal orientation shifted smoothly from column to column, like a minute hand creeping around a clock.

The geographical patterns didn't stop there, though. Further digging revealed that, just as cells worked together in columns, columns worked together in larger clusters, like a bundle of drinking straws. Each bundle had enough orientation columns to cover all 180 degrees of possible lines, from — to | and back to —. Each bundle also responded best to one eye, right or left. Hubel and Wiesel soon realized that one left-eye bundle plus one right-eye bundle — a "hypercolumn" — could detect any line with any orientation within one pixel of the visual field. Once again, this took years of work to firm up, but it turns out that no matter what lovely shape our eyes lock onto — the swirl of a nautilus shell, the curve of a hip — the brain determinedly breaks that form down into tiny line segments.

Eventually Hubel and Wiesel relieved their neckaches and got their apparatus turned the right way around, so that the clockwork kitties stared straight forward, toward a proper screen. And the discoveries just kept coming. Beyond simple line-detecting neurons, Hubel and Wiesel also discovered neurons that loved to track motion. Some of these neurons got all excited for up/down motion, others buzzed for left/right movement, and still others for diagonal action. And it turned out that these motion-detecting neurons outnumbered the simple line-detecting neurons. They outnumbered them by a lot, actually. This hinted at something that no one had ever suspected — that the brain tracks moving things more easily than still things. We have a built-in bias toward detecting action.

Why? Because it's probably more critical for animals to spot moving things (predators, prey, falling trees) than static things, which can wait. In fact, our vision is so biased toward movement that we don't technically see stationary objects at all. To see something stationary, our brains have to scribble our eyes very subtly over its surface. Experiments have even proved that if you artificially stabilize an image on the retina with a combination of special contact lenses and microelectronics, the image will vanish.

With these elements in place — Inouye's map of the visual cortex, plus knowledge of line detectors and motion detectors—scientists could finally describe the basics of animal vision. The most important point is that each hypercolumn can detect all possible movements for all possible lines within one visual pixel. (Hypercolumns also contain structures, called blobs, that detect color.) Overall, then, each one-millimeter-wide hypercolumn effectively functions as a tiny, autonomous eye, a setup reminiscent of the compound eyes of insects. The advantage of this pixilated system, besides acuity, is that we can store the instructions to create a hypercolumn just once in our DNA, then hit the repeat button over and over to cover the whole visual field.*

Some observers claimed that science learned more about vision during Hubel and Wiesel's two decades of collaboration than in the previous two centuries, and the duo shared a much-deserved Nobel Prize in 1981. But despite their importance, they took vision science only so far. Their hypercolumns broke the world down quite effectively into constituent lines and motion, but the world contains more than wriggling stick figures. Actually *recognizing* things, and summoning memories and emotions about them, requires more processing, in areas of the brain beyond the primary visual cortex.

~

Fittingly, the next advance in vision neuroscience—the "two streams" theory—appeared in 1982, just after Hubel and Wiesel won the Nobel. All five senses have primary processing areas in the brain, to break sensations down into constituent parts. All five senses also have so-called association areas, which analyze the parts and extract more sophisticated information. It just so happens with sight that, after the primary visual cortex gets a rough handle on something's shape and motion, the data get split into two streams for further processing. The *how/where* stream determines where something is located and how fast it's moving. This stream flows from the occipital lobes into the parietal lobes; it eventually pings the brain's movement centers, thereby allowing us to grab onto (or dodge) whatever we're tracking.

The *what* stream determines what something is. It flows off into the temporal lobes, and taps into the memories and emotions that make a jumble of sensations snap into recognition.

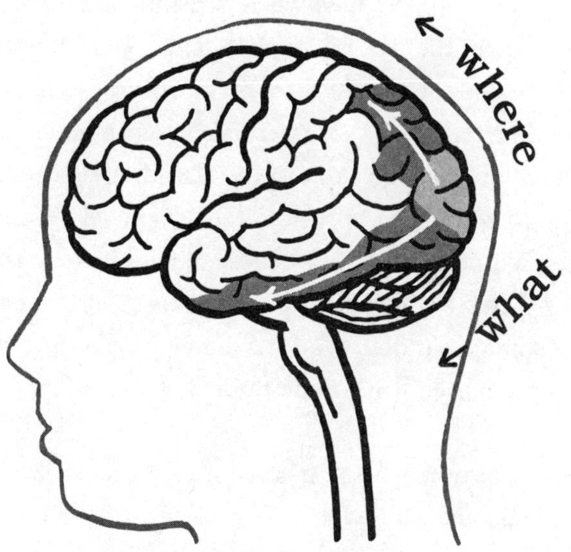

No one knows for sure how that snap takes place, but one good guess involves circuits of neurons firing in sync. At the beginning of the what stream, neurons are rather indiscriminate: they might fire for any horizontal line or any splash of red. But those early neurons feed their data into circuits farther upstream, and those upstream circuits are more picky. They might fire only for lines that are red *and* horizontal, for example. Still farther upstream, circuits might fire only for red horizontal lines with a metallic glint, and so on. Meanwhile, other neurons (working in parallel) will fire for clear glass lines at a certain angle or black rubber circles. Finally, when all these neurons throb at once, your brain remembers the pattern—red metal, glass, rubber—and says, *aha, a Corvette.** The brain also integrates, over a few tenths of a second, the Corvette's sound and texture and smell, to further aid in recognition. Overall, then, the process of recognition is smeared out among different parts of the brain, not localized in one spot. (Important note.*)

In everyday life, of course, we don't bother distinguishing between seeing a car (primary visual cortex), recognizing a car (what stream), and locating a car in space (how/where stream). We just look. And even inside the brain, the streams aren't independent: there's plenty of feedback and crosstalk, to make sure you reach for the right thing at the right time. Nevertheless, those steps are independent enough that the brain can stumble over any one of them, with disastrous results.

If the primary visual cortex suffers damage, people lose basic perceptual skills, a problem that becomes obvious when they draw things. If they sketch a smiley face, the eyes might end up outside the head. Tires might appear on top of cars. Some people can't even close a triangle or cross an X. This is the most devastating type of visual damage.

Damage to the how/where stream hinders the ability to locate objects in space: people whiff when they grab at things and constantly run into furniture. Even more dramatic, consider a fortysomething woman in Switzerland who suffered a parietal lobe stroke in 1978. All sense of movement disappeared for her, and life became a series of Polaroid snapshots, one every five seconds or so. While pouring tea, she saw the liquid freeze in midair like a winter waterfall. Next thing she knew, her cup overflowed. When crossing the street, she could see the cars fine, even read their license plates. But one moment they'd be far away, and the next they'd almost clip her. During conversations, people talked without moving their lips—everyone was a ventriloquist—and crowded rooms left her nauseous, because people appeared and reappeared around her like specters. She could still track movement through touch or sound, but all sense of visual motion vanished.

Finally, if the what stream malfunctions, people can pinpoint where objects are but can no longer tell one object from another. They cannot find a pen again if they put it down on a cluttered desk, and they're hopeless in parking lots at the mall. Bizarrely, though, they can still perceive surface details just fine. Ask them to copy a picture of a horse, a diamond ring, or a Gothic cathedral, and they'll render it immaculately—all without recognizing it. Some people can even draw objects from memory, but if shown their own drawings later, nothing registers. In general, these people retain their perceptual skills, since the primary visual cortex works, but the details never snap into recognition and identity eludes them.

Sometimes damage to the what stream is more selective, and rather than all objects, people fail to recognize only a narrow class of things. Many of these so-called category deficits arise after attacks of the herpes virus, the same bug that causes cold sores. Herpes means "creeping," and although it's normally harmless, the virus does sometimes go rogue and migrate up the olfactory nerves to the brain, where it ravages the temporal lobes. When this happens, neurons begin to fire in panic, and victims complain of funny smells and sounds; as more tissue dies, they suffer headaches, stiff necks, and seizures. Many fall into comas and die. Those patients who wake up again often have sharply focused brain damage, as sharply focused as if a Russian bullet had pierced them. And if just the right spot gets nicked, they might display a correspondingly sharp mental deficit. Most commonly, people lose the ability to recognize animals. Inanimate objects they recognize fine—strollers, tents, briefcases, umbrellas. But when shown any animal, even cats or dogs, they stare, mystified, as if looking at beasts dragged back from an alien zoo.

Loads of similar cases exist, some of which beggar belief. Contra the cases above, some herpes victims can recognize living things just fine, but not tools or man-made objects: cash registers become "harmonicas," mirrors become "chandeliers," darts fairy-tale transform into "feather dusters." (Scarily, one man with so-called object blindness continued to drive. He couldn't tell cars from buses from bicycles, but because his how/where stream still worked, he could detect motion, and simply steered clear of anything coming at him.) To get even more specific, some brain-damaged people can recognize objects and animals but not food. Others blank out only on certain categories of food, such as fruits and vegetables, while still others can name cuts of meat but not the animals they came from. "Color amnesiacs" cannot remember where lemons fit into the rainbow, nor whether blood and roses are of similar hues. One woman struggled with questions about, no kidding, the color of green beans and oranges.

Usually these "mind-blind" folks can identify things through another sense: let them touch a toothbrush or sniff an avocado, and it all comes back. Not always, however. One woman who couldn't recognize animals by sight also couldn't recognize animal sounds, even though she could identify inanimate objects via sound. She had difficulties with spatial dimensions, too, but again only with animals. She knew that tomatoes are bigger than peas, but couldn't remember whether goats are taller than raccoons. Along those lines, when scientists sketched out objects that looked like patent-office rejects (e.g., water pitchers with frying-pan handles), she spotted them as fakes. But when they drew polar bears with horse heads and other chimeras, she had no idea whether such things existed. For some reason, as soon as an animal was involved, her mind gummed up.

These pure category deficits, while rare, imply something important about the evolution of the human mind. Our ancestors spent a lot of time thinking about animals, whether furry, feathered, or scaly. The reason is obvious. We're animals ourselves, and the ability to recognize and pigeonhole our fellow creatures (as food, predators, companions, beasts of burden) gave our ancestors a big boost in the wild. Eventually, we probably developed specialized neural circuitry that took responsibility for analyzing animals, and when those circuits crap out, the entire category can slip clean out of people's minds. Our ancestors exploited fruits and vegetables, too, as well as small, tool-like objects. Probably not coincidentally, these are the two other categories of things that commonly disappear from people's mental repertoire. Our brains are natural taxonomists: we cannot help but recognize certain things as special. But the danger of specialized circuitry is that if the circuits go kaput, an entire class of things can go extinct mentally.

The way we catalogue the world teaches us something else about mind-brain evolution. I hesitate to even evoke the m-word, since it's such a contentious term. But after reading about fruit deficits and animal deficits and color deficits, it seems pretty clear that our brains do have *modules* on some level—semi-independent "organs" that do a specific mental task, and that can be wiped out without damaging the rest of the brain. Some neuroscientists go so far as to declare the entire brain a Rube Goldberg machine of modules that evolved independently, for different mental tasks, and that nature stuck together with gum and rubber bands. That "massive modularity" pushes things too far for some scientists: they see the mind-brain as a general problem solver, not a collection of specialized components. But most neuroscientists agree that, whether you call them modules or not, our minds do use specialized circuits for certain tasks, such as recognizing animals, recognizing edible plants, and recognizing faces.

~

In some sense we analyze faces the same way as other objects, by scribbling our eyes over the lines and shadows and contours we see, which causes certain ensembles of neurons to harmonize and hum. That said, analyzing faces requires more sophisticated brainware than analyzing other objects, both because social creatures like us need to read people's thoughts and feelings on their visages, and also because—let's face it—most people's features look pretty darn similar overall.

As with any mental faculty, many different patches of gray matter contribute to analyzing faces. But certain patches near the brain's south pole, like the fusiform face area, have special responsibilities. On brain scans the FFA lights up whenever people study faces, and disrupting it electrically causes faces to morph and stretch in funhouse ways. The most notable feature of the FFA is holistic processing. Instead of piecing together a face feature by feature—the way we seem to process regular objects—we read faces instantly, at a glance. In other words, a whole face is greater than the sum of the eyes and nose and lips in isolation.

To be sure, the FFA can light up in other circumstances. Ornithologists and auto aficionados and Westminster judges get lots of pings there when they study birds and cars and dogs, respectively. In other words, whenever we need to parse a narrow class of nearly identical things, our plastic brains might recruit the FFA to help out.

Still, the balance of evidence suggests we do have a specialized, if nonexclusive, face circuit. Even with object and animal aficionados, the FFA lights up strongest for faces. And beyond the FFA—which is just one component of a larger system—our brains also process faces in more complicated ways than other objects: we have circuits that light up only for certain emotional expressions or only when someone looks in a certain direction. Also unlike with cars or whatever, we constantly detect faces where they don't exist, in bathroom fixtures and tortillas and random piles of rocks on other planets (a tendency called pareidolia). Anytime we see two dark spots hovering above a quasi-horizontal line, we can't help but want to introduce ourselves. Seeing faces is mandatory.

At least for most people. The best evidence for a specialized face circuit comes from people who struggle to recognize faces because of damage to the FFA or faulty wiring there. Some face-blind people brush right past their dearest friends on the street without blinking. For birthday bashes, even their own, they might ask people to wear name tags; same for family reunions. To recognize people at all, they either listen to their voices; memorize how they walk; or scan for distinctive moles, scars, or haircuts. (The great portrait painter Chuck Close has severe face-blindness; this seems ironic at first, but his need to scrutinize faces probably enhances his talent.) Some face-blind people cannot even determine gender or age. A Welsh mining engineer who fell asleep after a few drinks and had a stroke woke up unable to tell his wife from his daughter. In another case, a lesion left an Englishman so bereft of face-recognition skills that he quit society and became a shepherd. After a few years he could tell most of his sheep apart by looks, but he never did get the hang of humans again.*

The selective *sparing* of face circuits can also reveal a lot. In 1988, in Toronto, a man named C.K. was struck by a car while jogging and suffered a closed head injury. Aside from some emotional outbursts and memory problems, he more or less recovered and eventually completed a master's degree in history with the help of a voice-activated computer. Still, one faculty never recovered: C.K. couldn't tell any inanimate objects apart, even food. His neurologists recalled taking him to a buffet and watching him shuffle around, bewildered. Everything looked like "differently colored blobs," and at the table he seemed to stab his fork at random and eat whatever he speared. At home he could no longer stage mock battles with his beloved toy soldier collections, since Greek and Roman and Assyrian armies all looked the same. He couldn't recognize body parts, either: more than once he tried giving the heave-ho to a strange pink thing poking out from his sheets—his foot. Yet for all this blundering about, C.K. proved a savant with faces and learned them readily. He even startled his neurologist in the shower at the gym once by helloing the doctor well before the doctor could place him.

Intrigued by the purity of his deficit, neuroscientists ran C.K. through a battery of facial-recognition tests. He proved he could recognize celebrities easily, even with parts of their faces blacked out; he could also recognize celebrities when scientists superimposed disguises (e.g., Groucho glasses) on them. He could instantly pick out all the faces in those find-the-hidden-picture puzzles that conceal faces in, say, a forest scene. He could recognize Bugs Bunny, Bart Simpson, and other cartoon characters, and recognized caricatures of Elvis, Bob Hope, and Michael Jackson. (Caricatures often send people's FFAs into a frenzy, because they exaggerate facial features. It's like face porn.) Most impressively, C.K. could scan a stranger's face in a photograph just once, then pick him out of a photographic lineup of near twins, even when the target was facing a different direction. On many of these tests C.K. scored higher than normal, control people did.

On the flip side, C.K. floundered on other tests. When shown upside-down faces, for instance, even faces he'd identified before, recognition always eluded him. Neuroscientists have long known that inverting any object hinders recognition, and that inverting faces hinders recognition even more than inverting animals, buildings, and other things. But while other people can usually puzzle through an upside-down face, C.K. was clueless. He couldn't even identify upside-down cartoon faces, something most people found laughably easy. Splintering or scrambling a face, by exploding it into parts, also flummoxed him. And when shown an Arcimboldo—those odd, sixteenth-century "portraits" pieced together from fruits and vegetables—C.K. rarely saw anything but the overall mien; he was oblivious to the pear noses and apple cheeks and green-bean eyelids that make the rest of us gasp.

C.K.'s troubles imply that the brain can normally recognize faces through two channels. There's the FFA circuit, which recognizes faces quickly and holistically. This system escaped damage in C.K. But the FFA circuit is picky: it needs to see the eyes hovering above the mouth and needs to detect rough symmetry, or it fails to engage. In that case a backup system should take over. This backup system is slower, and probably pieces upside-down or fractured faces together feature by feature. In other words, it treats the face more like an object. In fact, it probably employs our general object-recognition brainware—which explains why C.K. suddenly floundered, since his object-recognition skills hovered in the bottom percentile. Dehumanize a face—turn it into a mere object—and the face savant became face-blind.

~

Naturally, the same circuits you use to recognize people around you also light up when you recognize your own features in the mirror. But seeing your own face also stirs up deeper associations—it taps into your id, your ego, your sense of self. And it was this aspect of the self that the facial wounds of World War I so threatened.

The study of facial disfigurement really took off in the twentieth century, and not only because of modern war. The rise of handguns and especially cars produced plenty of accidents among civilians. Surprisingly, though, in all groups studied, many disfigured people rebounded quite well: even some of the most severely injured showed few psychological hang-ups. Like the *mutilés* who married their nurses, these people tended to brush the disfigurement aside and keep living life. Some also joked about their scars when they caught people staring, mentioning a botched bear-wrestling career or saying, "God hit me with a frying pan."

Still, many victims reacted more predictably. They showed symptoms of mourning at first, grieving for their faces as for the dead. And they remained isolated long after their physical injuries had healed, suffering the gapes and double takes in silence. Years after the injury, a few were still startled by their reflections in plate-glass windows. A self-image is hard to let go of.

In the past decade psychologists have expanded their understanding of facial trauma by studying a new group of patients—the recipients of face transplants. A face transplant involves just what it sounds like, the surgical transfer of a nose, lips, cheeks, and other tissues from a dead person to a living one. In this way it combines the heroic reconstructive surgery of World War I with the lifelike masks of Anna Coleman Ladd and others. What's more, because a face transplant involves a living mask, a mask that can speak and express emotions, psychologists could finally probe the question that Ladd's work evoked so long ago: would the brain accept a new face as its own?

The first recipient of a face transplant, a thirty-eight-year-old Frenchwoman named Isabelle Dinoire, swallowed a mouthful of sleeping pills in May 2005 after an argument with her daughter. She didn't expect to wake up again but did. Groggy, she put a cigarette in her mouth and found it wouldn't stay. That's when she noticed the pools of blood: her Labrador retriever had mauled her while she slept. Dinoire staggered to a mirror. Tousled, dirty-blond hair still ringed her face, but the dog had gnawed her nose down to two skeletal holes, and no lips covered her teeth or gums. Although emergency care stabilized her, in later months Dinoire became a recluse, hiding behind a surgical mask at all times.

In the years leading up to Dinoire's injury, the medical world had worked itself into a froth over the ethics of face transplants. A few fear-mongering doctors actually suggested that donors' families might start stalking transplant recipients, or that a black market in beautiful faces would spring up. Some activists proposed banning even the *discussion* of face transplants, to spare the feelings of the already disfigured. Less hysterical types opposed the surgery on medical grounds. Transplanting skin provokes an especially strong immune response, so transplant recipients would have to take heavy-duty immunosuppressants, increasing their risk for catching many diseases and probably shortening their lives.

Nevertheless, other doctors pursued the idea. They cited surveys suggesting that people would indeed trade many years of life to restore a damaged face. Surgeons favoring facial transplants also pointed out that naysayers had sown similar fears of identity crises before the first heart transplants, and none of those had come to pass. Doctors emphasized the limits of alternative treatments as well. Plastic surgeons could do clever things like fashion a new nose from a toe (really), but it often looked terrible and obviously didn't function the same way. There's just no substitute for facial tissue.

In examining the risks of face transplants, doctors turned to whatever approximations they could find. To determine whether the new face would look more like the donor (who would supply the skin and cartilage) or the recipient (who would supply the underlying bone structure), surgeons swapped faces on cadavers, then asked volunteers to judge before-and-after photographs. They concluded that (aside from certain features, like eyebrows) the face would look different than both the donor and recipient. It would be a new, unique face. Doctors examined the outcomes of other radical transplants as well, like tongue, larynx, and especially hand transplants. As with face transplants, hand transplants involved multiple kinds of tissue, so the demands on the patient's immune system would be similar. Hand transplants also proved that the brain could integrate neurologically demanding tissues pretty easily. (It probably helps that, as with faces, we have specialized neurons that fire only in response to viewing hands—a legacy of hand gestures in prelanguage communication.*)

Doctors also evaluated the psychology of transplants. First and foremost, people needed to accept the foreign tissue as part of them. With hands, doctors made sure to correct any Freudian slips, forcing patients to refer to "my hand," not "*the* hand" in conversation. Doctors also emphasized the need to use the hands in daily activities, the more intimate the better: while the surgeons on one transplant team scowled to see a patient nervously chewing the nails on his new hand, his psychologists rejoiced—you don't bite someone else's fingernails. Unfortunately, those psychological safeguards didn't always work. The first hand transplant, in 1998, for one Clint Hallam, had gone quite well surgically, and Hallam had felt sensation creeping back into his new hand at a rate of a few millimeters per day. But after twenty-nine months Hallam stopped taking immunosuppressants, saying that the hand now creeped him out. His immune system attacked it, and doctors had to amputate.

If something went wrong with a face transplant, amputation wasn't an option. Nevertheless, French surgeons—who tempted fate by comparing themselves to Copernicus, Galileo, and Edmund Hillary—pushed ahead in 2005 with Isabelle Dinoire, the woman whose Labrador had mauled her. They picked Dinoire partly because she'd lost "only" her nose, lips, and chin (the facial triangle), which made for an easier surgery. A suitable donor turned up in November 2005, when a forty-six-year-old woman in a nearby town tried to hang herself and ended up brain-dead. She matched Dinoire in age, blood type, and skin tone, and Dinoire's surgeons rushed into action. They spent hours "recovering" the hanged woman's face—peeling away her skin and connective tissue along with blood vessels and nerves, leaving only a red mask of muscle behind. The transfer onto Dinoire then took the better part of a day.

During recovery, Dinoire's new face swelled frightfully, and on day eighteen her body nearly rejected it. Meanwhile the media got pretty frenzied; British tabloids even outed the identity of the brain-dead donor. But Dinoire recovered better than anyone could have hoped. She was eating with her new lips within a week and talking shortly thereafter. Hot and cold sensation returned within a few months, as did most movement. Most important, she started leaving the house again, resuming her social life and meeting new people. The one facial movement that lagged was smiling—at ten months she could only half smile, like a stroke victim. But by fourteen months she could smile fully again. She had reason to.

Chinese surgeons performed the second face transplant in April 2006, and more soon followed, with remarkable results. Many patients could speak, eat, and drink by day four. Sensation usually returned within a few months. And brain scans revealed that their faces came back "online" quickly, much more quickly than hands did. (Patients in fact got a kick out of watching their once-dormant face territories "wake up" on the scans.) The psychological adjustment usually went smoothly as well. It seemed to help that, unlike with a hand, you don't have to look at your face constantly. And when people did look in a mirror, they found it easy to accept the reflection. It wasn't the old "them," certainly. But the underlying bone structure was enough to evoke a feeling of "me, that's *me*" in the mirror.

Bolstered by these early successes, a few teams have now performed the more demanding full-face transplant. One early recipient, the third full-facial, was Dallas Wiens. In November 2008 the twenty-three-year-old Wiens was painting some structures on the roof of a Fort Worth, Texas, church when he accidentally steered his hydraulic lift into some power lines. The air around his head reportedly glowed blue for fifteen seconds, and the current running through his face melted it into a blank mask, one writer noted, not unlike "Mr. Potato-Head without the features."* In March 2011 Wiens got a replacement. The new face arrived in a blue cooler in a slurry of ice water; it was the size and thickness of medium pizza dough when unfurled.

Surgeons first hooked the donor face up to Wiens's blood supply through his carotid arteries. This took some creative suturing, since the donor had cigar-sized carotids, while Wiens's vessels (which had atrophied) looked like drinking straws. The transplant team felt enormous relief when the face started to flush pink, a sign that it was taking blood. In all, the surgery ran seventeen hours, during which time Wiens's new face smirked, winked, and grimaced as surgeons manipulated it to reattach various nerves and muscles. Afterward, doctors rolled him into the ICU to see if Wiens would be able to smirk, wink, and grimace on his own.

When Wiens awoke, he felt his new, swollen face pressing down hard, like a lead mask. He could breathe only through a tube in his trachea. But all the discomfort seemed worth it a few days later. In a moment so mundane it's poignant, he found he could finally smell food again. Lasagna. Touch sensation returned not long after, and he felt, really felt, his daughter's kiss for the first time in years. Wiens even began dreaming of himself with his new face. These were moments the World War I masks, even the most artistic, could never replicate.

As with hand transplants, doctors found that the more patients used their transplanted faces—shaving, smiling, applying makeup, smooching, getting smooched—the more they accepted the new faces as theirs, regardless of what they looked like. Humans do rely on vision to an extraordinary degree, and our visual circuits occupy far more brain territory than other sensory circuits. It's no surprise that looks are so tied up with our sense of self. Ultimately, though, one important truth of neuroscience is that the brain constructs our sense of self from more than mere looks. As we'll see later, our sense of self also draws on our emotional core and our memories and our personal narratives of our lives. The earliest face transplants took place in 2005, so their long-term medical viability remains unknown. But psychologically, at least, they've succeeded: the brain will indeed accept a new mien in the mirror—in part because it's only a mien, a covering. As one observer noted, "If a face transplant demonstrates anything about what it means to be human, it may be that we are less superficial than we imagine."

PART III

BODY AND BRAIN

CHAPTER FIVE

The Brain's Motor

Now that we've learned about some internal brain structures, it's time to explore how the brain interacts with the outside world. It does so primarily through movement, which involves transmitting messages to the body via nerves.

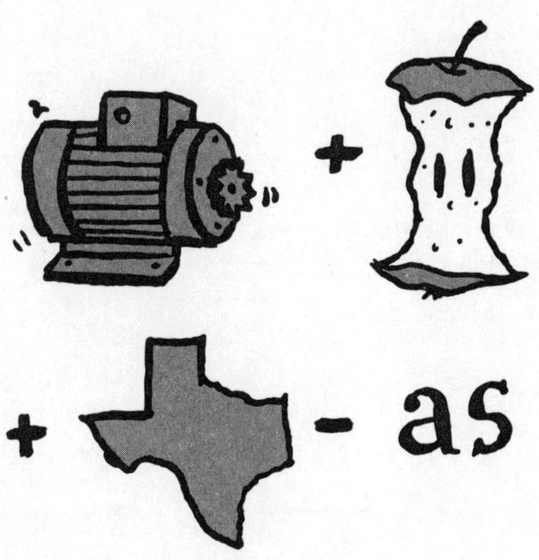

George Dedlow. The dimes and quarters and silver dollars trickling into Philadelphia's "Stump Hospital" often came with notes of sympathy for George Dedlow. Every man crowding around the hospital's front door wanted to tip his hat to, and every woman blow a kiss at, George Dedlow. The hospital superintendent pleaded ignorance, but well-wishers never tired of asking about Captain George Dedlow.

The cover story of the July 1866 issue of *The Atlantic Monthly* was "The Case of George Dedlow," one of the saddest of the Civil War's many sad tales. In the introduction Dedlow claimed that he'd originally tried to publish his report in a proper medical journal, and after a number of rejections had transformed the piece into a personal narrative. The action began with Dedlow joining the 10th Indiana Volunteers as assistant surgeon in 1861, despite having completed just half of medical school. The U.S. military was so desperate for surgeons then — it had just 113, a small fraction of the 11,000 that both armies would require during the war — that most units snatched up even tyros like him.

One night in 1862, Dedlow writes, while stationed near a malaria-ridden marshland south of Nashville, he received orders to sneak through twenty miles of enemy lines and secure some quinine. Seventeen miles in, he stumbled into an ambush and got shot in both arms — in his left biceps and right shoulder — and passed out. He woke to find the rebels, like centurions at the cross, drawing lots for his hat, watch, and boots. They eventually dumped him into a medical cart, which rattled him 250 miles south to an Atlanta hospital. His right arm throbbed the whole journey, burning as if it were being held near flames; he found relief only by dousing it with water. The burning continued for six weeks, and the pain became so acute that when his doctor suggested amputating the arm, Dedlow agreed despite the lack of ether.

After recovering, Dedlow was exchanged for a Southern captive.

Rather than return home, the one-armed doctor took just thirty days of furlough and rejoined his unit. The Indiana boys ended up in Tennessee again, and once again Tennessee did not treat them well. During one of the bloodiest battles in U.S. history, near Chickamauga Creek, Dedlow's unit got caught in intense crossfire while scampering up a hill. Clouds of gun smoke enveloped them, punctured by red lightning and rifle thunder. This time Dedlow got shot through both legs, one of the battle's 30,000 casualties.

He awoke beneath a tree, in shock, with two shattered femurs. Orderlies gave him brandy and cut away his pantaloons while two surgeons—wearing navy-blue uniforms with green sashes about the waist—bent to examine him. They grimaced and walked away, damning him by triage. Sometime later, though, Dedlow felt a towel against his nose, then inhaled the fruity chemical pinch of chloroform. Two other surgeons had returned, and although Dedlow didn't realize it, they'd decided to amputate both legs right there in the field.

Confederate surgeons usually performed "circular" amputations. They made a 360-degree cut through the skin, then scrunched it up like a shirt cuff. After sawing through the muscle and bone, they inched the skin back down to wrap the stump. This method led to less scarring and infection. Union surgeons preferred "flap" amputations: doctors left two flaps of flesh hanging beside the wound to fold over after they'd sawed through. This method was quicker and provided a more comfortable stump for prosthetics. Altogether, surgeons lopped off 60,000 fingers, toes, hands, feet, and limbs during the war. (In Louisa May Alcott's *Hospital Sketches,* one Union soldier proclaims, "Lord!, what a scramble there'll be for arms and legs, when we old boys come out of our graves on the Judgment Day.") A typical amputation lasted maybe four minutes, and on the worst days a surgeon might do a hundred—some in fields, some in barns, stables, or churches, some on nothing but a plank suspended between two barrels. In borderline cases surgeons erred on the side of amputating, since the mortality rate for compound fractures was abysmal. Not that the mortality rate for amputations was good. Sixty-two percent of double thigh amputees died.

To his later sorrow, Dedlow woke up after his double thigh amputation. But it was at this very moment—in his fog, before he realized what had happened—that Dedlow's tale swerved and began to transcend a typical soldier's sob story. For despite the operation, Dedlow awoke with cramps in both calves.

He hailed a hospital attendant, gasping, "Rub my left calf."

"Calf? You ain't none," the attendant answered. "It's took off."

"I know better. I have pain in both legs."

"Wall, I never. You ain't got nary leg." With this, Dedlow recalled, "he threw off the covers and, to my horror, showed me..."

Faintly, Dedlow dismissed him. He lay back, ill, probably wondering if he'd gone mad. But damn it, he'd felt the cramps in both legs. They'd *felt* intact.

Soon another tragedy befell him. Dedlow's left arm had never quite healed right after the ambush near Nashville and continued to weep pus. Now, in the dirty recovery ward, the arm contracted "hospital gangrene," an aggressive disease that could eat away flesh at rates of a half inch per hour. Almost half of all victims died in their beds, and against his better judgment, Dedlow let his doctors save his life by amputating his last remaining limb. He awoke to find himself, he later sighed, a diminished thing, more "larval" than human.

In 1864, Dedlow was transferred to Philadelphia's South Street Hospital—nicknamed "Stump Hospital" for all the amputees limping through its corridors. But even within Stump, Dedlow's helplessness set him apart: orderlies had to dress him every morning, had to drag him to the toilet at all hours, had to blow his nose and scratch his every itch. Virtually sedentary—orderlies had to carry him everywhere in a chair—he needed almost no sleep, and his heart beat just forty-five times per minute. With so little body to nourish, he could barely finish the meals that orderlies had to feed him bite by bite.

Yet he could somehow still feel that missing four-fifths of himself—still feel pain in his invisible fingers, still feel his invisible toes twitching. "Often at night I would try with one lost hand to grope for the other," he recalled, but the ghosts always eluded him. Out of curiosity he interviewed other Stump inmates and discovered that they felt similar sensations—stabbing, cramping, itching—in their missing limbs. Indeed, the ungodly aches in their ghost arms and ghost legs often made their missing limbs more insistent and intrusive than their real ones.

Dedlow didn't know what to make of this phenomenon until, a few depressing months later, he met a fellow invalid, a sergeant with washed-out blue eyes and sandy whiskers. They struck up a conversation about spiritualism and communicating with departed souls. Dedlow scoffed, but the sergeant talked him into attending a séance the next day. There, after some preliminary mumbo jumbo, the mediums started summoning up people's dead children and late spouses—a trick that often reduced the participants to hysterics. The mediums also relayed messages from the beyond, Ouija board–style, by pointing to letters on an alphabet card. They then listened for a confirmatory knock (spirits can knock, apparently) upon reaching the correct letter. Eventually a wan medium with bright red lips named Sister Euphemia approached Dedlow. She asked him to silently summon to mind whomever he wanted to see. All at once, Dedlow says, he got a "wild idea." A moment later, when Euphemia asked if Dedlow's guests were present, two knocks sounded. When Euphemia asked their names, they tapped out, cryptically, "United States Army Medical Museum, Nos. 3486, 3487."

Euphemia frowned, but Dedlow, a war surgeon, understood. As reported by Walt Whitman (and many others who couldn't shake the image from their minds), hospitals routinely piled all their amputated limbs outside their doors, forming cairns of legs, arms, and hands. Rather than bury them, though, the army packed the flesh into barrels of whiskey and shipped them to the Army Medical Museum, which catalogued them for future study. Dedlow's legs were apparently numbers 3486 and 3487, and per his wish, Euphemia had summoned them to the séance.

At this point the story swerved again. Dedlow suddenly cried out, then began to rise in his chair. He reported feeling his ghost legs beneath him, reattaching themselves to his femurs. A moment later his torso rose, and he began staggering forward. He felt unsteady at first—after all, he noted, his legs had been soaking in booze. But he crossed half the room before they dematerialized, at which point he collapsed.

Here Dedlow ended his story abruptly. Rather than cheer him, the brush with the other side only reminded him of what he'd lost, and he felt even more diminished. As he told the orderly transcribing his story, for any man "to lose any part [of himself] must lessen...his own existence." He concluded, "I am not a happy fraction of a man."

Although rejected by medical journals, "The Case of George Dedlow" panged people—pierced them in a way that an academic paper never could have. The Civil War had maimed and disfigured hundreds of thousands of men. Nearly everyone had a brother or uncle or cousin whose wounds had never set right. Moreover, as the first well-photographed war, the Civil War branded the country's psyche with indelible images, of stumps and naked wounds and holes where there shouldn't be holes. These macabre photographs, in museums, in magazines, were in some ways the heir to Vesalius's *Fabrica.* Except they didn't celebrate the human form so much as catalogue its destruction.

And yet for all their power, these images of broken men remained silent—until George Dedlow gave them voice. His story spoke for every misshapen soldier in every village square, for every sobbing wreck in every parish pew, for every amputee whose ghostly limb made him scream out in the night.

So from far and wide that summer of 1866, donations arrived in Philadelphia for Captain Dedlow. Crowds even gathered around Stump's front door, pleading to meet their hero—and were stunned to hear that Dedlow didn't exist. With much regret, the hospital superintendent told the throngs that there was no George Dedlow among his patients. Nor could he find any George Dedlow in the hospital archives. For that matter, the military had searched its records and could find no cases, anywhere, of any quadruple amputees. The tale in *The Atlantic Monthly*, the superintendent explained, was fiction. The only authentic thing about it was Dedlow's disorder, a disorder medicine had never taken seriously before. The only real detail was, paradoxically, the phantom limbs.

~

For as long as human beings have waged wars, surgeons have lopped off limbs—although until recently soldiers rarely lived to speak of the experience. Similar to his reforms with treating gunshot wounds, Ambroise Paré convinced surgeons in the 1500s not to cauterize fresh stumps by dunking them in boiling oil or sulfuric acid. Instead Paré promoted ligation, which involved tying off the severed ends of arteries and sewing the stump shut. This greatly reduced blood loss and infection (not to mention agony), and meant that amputees finally had a decent chance of surviving. Paré became so confident of their survival, in fact, that he started designing fake limbs for them, some of which, thanks to gears and springs, actually moved. (His line of substitute ears, noses, and penises remained immobile, however.)

Not surprisingly, the first stray references to phantom limbs appeared in Paré's writing, and they quickly became an object of fascination for philosophers. René Descartes dabbled in neuroscience at times—he famously declared the pineal gland,* a pea-sized nugget of flesh just north of the spinal cord, the earthly vessel of the human soul—but he also ruminated on the implications of phantom limbs. One story, about a girl who lost her hand to gangrene but woke up moaning about the pain there, especially shook him. This and related stories "destroyed the faith I had in my senses," he wrote—to the point that he stopped trusting the senses as a sure route to knowledge. From there it was but a small step to *cogito ergo sum*, a declaration that he had faith only in his powers of reasoning.

British naval hero Horatio Nelson also leapt from phantom limbs to metaphysics. During the biggest blunder of his career—an attack on Tenerife, in the Canary Islands, in 1797—a musket ball shredded his right shoulder, and a surgeon had to hack it off in the dim hold of a rolling ship. For years afterward Nelson felt his phantom fingers digging into his phantom palm, causing excruciating pain. He actually took succor from this, citing it as "direct proof" that the soul existed. For if the spirit of an arm can survive annihilation, why not the rest of a man, too?

The physician Erasmus Darwin (grandfather of Charles), philosopher Moses Mendelssohn (grandfather of Felix), and writer Herman Melville (in *Moby-Dick*) all touched on phantoms as well. But the first clear clinical account of phantom limbs—he even coined the term—came from Civil War doctor Silas Weir Mitchell.

Weir Mitchell—he hated the name Silas—grew up a dreamy lad in Philadelphia. He suffered from phantasmagoric nightmares after hearing about the "holy ghost" in church, and he dabbled in both poetry and science. He especially loved the bright, pretty concoctions

Neurologist Silas Weir Mitchell.

his father, a doctor, would conjure up in his private chemical laboratory. Mitchell eventually decided to enter medical school—over the objections of his old man, who thought he wouldn't stick it out. Mitchell did, and even did rigorous medical research on snake venom before settling into private practice in Philadelphia in the 1850s.

Despite hating slavery, Mitchell didn't take the outbreak of the Civil War that seriously. Like many Americans, north and south, he assumed his side would whip the other in short order, and that would be that. He soon realized his mistake and became a contract military doctor. After a few months of making rounds to different military hospitals, Mitchell discovered he had a knack for neurological cases, cases most doctors loathed, even feared. So as the bodies kept piling up—Philadelphia's patient population reached 25,000 during the war—he helped found a neurological research center, Turner's Lane Hospital, on a dirt road outside Philadelphia in 1863.

One patient called Turner's Lane "a hell of pain"—a fair assessment, although this was partly by design. The military arranged for most cases of neurological trauma to end up there, and Mitchell preferred to trade away "easy" cases to other hospitals for more challenging ones, swapping convalescents with simple stomach wounds for thrashing epileptics and howling infantrymen with shattered skulls. Turner's, then, became the hospital of last resort, and although many of his patients never recovered, Mitchell found the work rewarding. He became an expert on nerve damage and especially on phantom limbs, since the Civil War produced amputees on an unprecedented scale.

A few months after Turner's Lane opened, Mitchell rushed over to the Battle of Gettysburg, where he saw for himself why the Civil War left so many limbless. Before the 1860s most soldiers used muskets. Muskets loaded from the front and they loaded quickly, since the bullets had smaller diameters than the barrels. This gap between bullet and barrel, however, produced swirling air currents that spun the bullet chaotically as it zipped down the barrel's length. As a result, the bullet curved when it emerged from the muzzle, like a doctored baseball. This made aiming all but pointless: as one Revolutionary War veteran sighed, "[when] firing... at two hundred yards with a common musket, you might as well just fire at the moon."

The other common type of military gun, the rifle, had the opposite problem: it was accurate—soldiers could plug a turkey's wattle at several hundred yards—but slow. The key to the rifle's dead aim was the inner barrel, which had tight, spiraling grooves running along its length; these grooves spun a bullet aerodynamically, like a football. For the grooves to work, however, the bullet and barrel had to be in close contact. This required bullets and barrels of basically the same diameter—which made them a bitch to load. Soldiers had to ram the bullets down the barrels inch by inch with rods, a laborious process that led to lots of jamming and swearing.

A few enterprising soldiers finally combined the best of rifles and muskets in the 1800s. An Englishman stationed in India noticed that warriors often tied hollow lotus seeds to their blow darts. When fired with a puff, the seeds ballooned outward and hugged the peashooters' barrels as they moved forward, much like a rifle. Inspired, the Englishman invented a metal bullet that had a hollow cavity, and in 1847 a Frenchman named Claude-Étienne Minié (*min-YAY*) greatly improved the design. Minié's bullets were smaller than a rifle's barrel, so they loaded quickly. At the same time, like the lotus seeds, they expanded when fired (from a punch of hot gas) and hugged the barrel's grooves as they hurtled forward—making the guns uncannily accurate. Worse, because the bullets had to expand, Minié made them out of soft, pliable lead. This meant that, unlike those hard Russian bullets forty years later, Minié bullets deformed upon impact, widening into blobs and shredding tissue instead of passing clean through. The result was an awesome killing machine. Based on its accuracy, its rate of fire, and the likelihood of a gaping wound, historians later rated the Minié bullet/rifle combination as three times deadlier than any gun that had ever existed. And those soldiers who didn't die had their limbs shattered beyond repair.

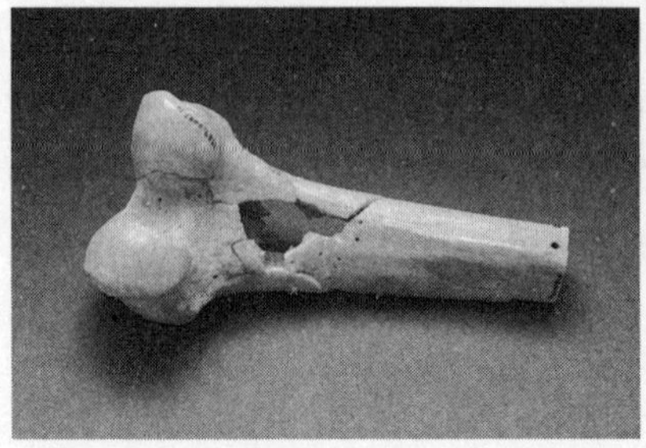

Left: A femur bone, shattered and amputated after being struck with a Minié bullet. Right: Minié bullets, made of lead. (National Library of Medicine)

In 1855 the secretary of war, Jefferson Davis, selected the Minié/rifle combo as the U.S. military's official arms and ammo. Six years later, as president of the Confederacy, Davis no doubt rued his earlier enthusiasm. Manufacturers started churning out untold numbers of cheap Minié bullets—which soldiers called "minnie balls"—and factories in the North especially started stamping out millions of Minié-compatible rifles, which butchered boys almost from sea to shining sea. The guns weighed ten pounds, cost $15 ($210 in today's money), and measured about five feet long. They also had an eighteen-inch bayonet, which was risible, since this gun more or less rendered the bayonet a foolish relic: rarely could soldiers get close enough to plunge one in anymore. (Mitchell once estimated that mule kicks hurt more soldiers during the Civil War than bayonets.) The Minié bullet also pushed cannons far back behind the infantry lines and greatly diminished the power of the mounted cavalry charge, since horses were even easier to pick off than humans. By some estimates Miniés killed 90 percent of the soldiers who died on the battlefield.

Unfortunately, many Civil War commanders—steeped in antiquated tactics and drenched in the romance of Napoleonic charges—never adjusted to the new reality. Most notoriously, on the day Mitchell arrived at Gettysburg, some 12,500 Confederate soldiers stormed a stone fence held by the Union. Pickett's Charge. Among other troops, soldiers with piles of minnie balls were waiting, and they pulped the guts and pulverized the bones of the chargers up and down the line.

An injured soldier might languish for days before a stretcher team or ambulance wagon lugged him to a clinic. There, he might wait hours more until a surgeon in a bloody apron appeared, a knife between his teeth. The surgeon would probe the wound with fingers still crimson from the last patient, and if he decided to amputate, one assistant knocked the patient out with chloroform or ether, another put the limb into a headlock, and a third got ready to clamp the arteries. Four minutes later, the limb fell. The surgeon hollered "Next!" and walked on. This work might continue all day—one Kentucky surgeon remembered his fingernails getting soft from absorbing so much blood—and fresh graves ringed every hospital.* Walt Whitman recalled the crude tombstones, mere "barrel-staves or broken boards stuck in the dirt."

After Gettysburg, Mitchell returned to Philadelphia to deal with the deluge of casualties. And although he continued his private practice (military work paid just $80 per month), he spent the better part of most days at Turner's Lane, arriving at 7 a.m. for an hour of rounds, then returning around 3 p.m. and often staying until midnight. He spent hours writing up case reports as well—an illuminating experience. His early research training had emphasized rigor and data, but Mitchell found he couldn't capture these cases with numbers and charts alone. Only narrative accounts could get at what injured soldiers really felt. The narratives affected him so profoundly, in fact, that in later years he began writing novels about his experiences, and drew on these case reports for inspiration.

Mitchell did his best and most original research on phantom limbs. Before his time, relatively few people admitted to them, since they risked being pegged as loopy. A more sympathetic Mitchell determined that 95 percent of his amputees experienced ghost limbs. Interestingly, though, the distribution of phantoms wasn't equal: patients felt upper-body phantoms more vividly than lower-body phantoms, and felt phantoms in the hands, fingers, and toes more acutely than phantoms in the legs or shoulders. And while most men's phantoms were paralyzed—frozen into one position—some soldiers could still "move" their phantoms voluntarily. One man would raise his phantom arm instinctively, to grab for his hat, whenever a gust kicked up. Another man missing a leg kept waking up at midnight to use the privy; groggy, he'd swing the phantom leg onto the floor and tumble.

Mitchell also probed phantom pain. Cramps or sciatica might race up and down the phantom, in waves lasting a few minutes. Less acute, but possibly more maddening, people's phantom fingers or feet would start itching—itches impossible to scratch. Stress often exacerbated the discomfort, as did yawning, coughing, and urinating. Perhaps most important, Mitchell determined that if a soldier had felt a specific pain right before his amputation—like fingernails digging into his palm, a common result of muscle spasms—that same pain often got "stamped" into his nerves, and would persist for years afterward in the phantom.

To explain where phantoms came from, Mitchell suggested a few interrelated theories. His patients' stumps often had raised growths on them where the underlying nerves had been severed. These "buttons" proved quite sensitive to the touch; they prevented many men from wearing prosthetics. Mitchell deduced from this touchiness that the nerves beneath must still be active—and still pinging the brain. As a result, part of the brain didn't "know" the limb had gone AWOL. As further proof, Mitchell cited a case where he'd actually resurrected a patient's phantom. This man had stopped feeling his phantom arm years before (as sometimes happened), but when Mitchell applied an electric current to the stump buttons, the man felt his former wrist and fingers suddenly materialize at the end of his stump—exactly as George Dedlow had at the séance. "Oh, the hand, the hand!" the man hollered. This indicated that the brain did indeed take cues from the stump.

Mitchell also implicated the brain itself in phantom limbs, a crucial development. Many a veteran, despite losing his dominant hand decades before, kept eating meals and writing letters with that hand in his dreams. Unlike stump irritation, this was a purely mental phenomenon and therefore must have its origins within the brain. Even more arresting, Mitchell discovered that some people who'd lost a hand or leg in infancy, and therefore had no memory of it, nevertheless experienced phantoms. Mitchell concluded from these cases that the brain must contain a permanent mental representation of the full body—a four-limbed "scaffold" stubbornly resistant to amputation. The brain's private metaphysics, then, trumped physical reality.

Later work by other scientists confirmed and built upon Mitchell's insights. For instance, Mitchell focused on how preamputation pain or paralysis can carry over into the phantom, but later scientists found that less pernicious sensations can be stamped onto the ghost as well. Some amputees feel phantom wedding bands and Rolexes, and people whose arthritic knees or knuckles allowed them to sense impending thunderstorms can often pull off the same trick with their phantoms. Moreover, neuroscientists have confirmed Mitchell's guess that the brain contains a hardwired scaffold of the full body, since children born without arms or legs sometimes still feel phantoms. One girl born without forearms did arithmetic in school on her phantom fingers.

Doctors have also catalogued phantoms in brave new places. Dental extractions can produce phantom teeth. Hysterectomies can produce phantom menstrual cramps and labor pains. After colorectal procedures, people might feel phantom hemorrhoids and bowel movements and rumbling phantom flatulence. There are also phantom penises, complete with phantom erections. Most phantom penises arise after penile cancer or accidents with shrapnel that most of us prefer not to think about. But unlike phantom limbs—which are often frozen into claws, and excruciating—most men find a phantom penis pleasurable. And they're so realistic that even decades after the penis is shorn, some men still walk a little funny when they get aroused. Heck, some men's phantom penises lead to real orgasms. All this showed that quite a few sensations and emotions in the brain can be tied up with phantoms.* The work furthermore helped shift the focus of phantom limbs away from the stump and toward the brain itself.

~

Although Mitchell made phantom limbs an object of legitimate scientific study, this knowledge didn't readily translate into treatments. For most of the twentieth century, in fact, no different than in Mitchell's day, doctors merely fit amputees with prosthetics and, if the phantom pain got bad, plied them with opiates. But in the 1990s phantom research went through a renaissance, as neuroscientists realized that it provides a unique glimpse into the brain's movement centers and especially into brain plasticity.

The brain's primary movement center is the motor cortex, a strip of gray matter that starts near your ears and runs to the top of your head. It sends out the commands that spur the spinal cord to move your muscles. On its own, however, the motor cortex can produce only crude movements, like kicks and lunges. Think of a bucking

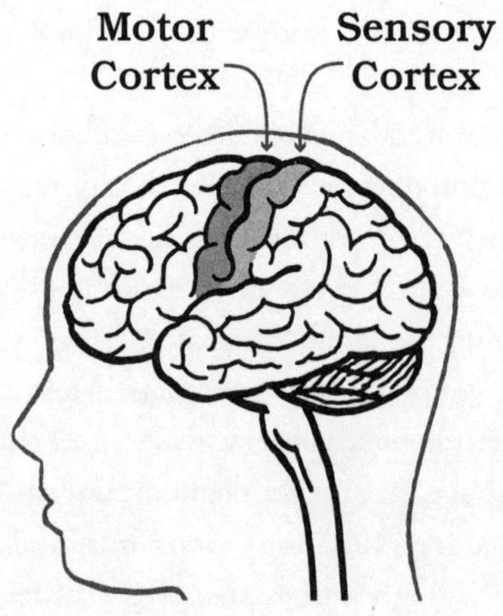

bronco—powerful, but lacking grace. Synchronized movements actually arise from two adjacent regions, the premotor cortex and supplementary motor area. In essence, these two regions coordinate simple movements into something more balletic. To change analogies, they play the motor cortex like a piano, pressing different areas in quick succession to produce complex chords and arpeggios of movement—walking, for instance, requires different muscle groups to contract with a precise amount of force at different moments. Toddlers stumble so much in part because their brains still hit false notes.

To execute a complicated movement, the motor areas also need feedback from the muscles at each stage, to ensure that their commands have been carried out properly. Much of this feedback is provided by the somatosensory cortex, the brain's tactile center. You can think about the somatosensory cortex as the motor cortex's twin. Like the motor cortex, it's a thin, vertical strip; they in fact lie right next to each other in the brain, like parallel pieces of bacon. Both strips are also organized the same way, body part by body part; that is, each strip has a hand region, a leg region, a lips region, and so on. In effect, then, the motor cortex and somatosensory cortex each contain a "body map," with each body part having its own territory.

In some ways this body map is straightforward; in other ways it's not. For example, just like on your body, the map's hand region lies right next to the arm region, which lies right next to the shoulder region, and so on. But in other spots, the topography is scrambled. In particular, the hand territory also borders the face territory, even though the hand itself doesn't border the face. Just as randomly, your foot territory nestles against the crotchal region.

The brain's body maps also contain another counterintuitive feature. Despite what you might think, big body parts don't need big patches of gray matter to function. Legs, for instance, although powerful, don't require complicated instructions to jump or kick, and they're not very sensitive to touch, either. As a result, these big burly parts get by with minuscule, Luxemburg-sized territories on the touch and movement maps. The lips, tongue, and fingers, meanwhile, engage in intricate movements such as speaking and handling tools, and therefore need Siberia-sized tracts of neurons. Some body parts, in other words, are magnified on the maps. (This explains why amputee soldiers felt missing fingers more than missing hands, and missing hands more than missing arms: our brains pay more attention to fine-motor structures.)

With all that in mind, consider what happens when a hand is amputated. First, a huge territory on the brain map goes black. It would be like watching the United States from space at night, with all its patches of sprawling suburban illumination, and seeing the power grid in Chicago fail. The key point, however, is that this spot doesn't stay blank. Because the brain is plastic, adjacent areas can colonize the hand region and use its neurons for their own ends. With a missing hand, it's usually the resource-hungry face territory that encroaches.

This encroachment happens quickly, sometimes within days, and happens over long neural distances, up to an inch. For these reasons scientists suspect that colonization cannot simply involve new neuron tendrils sprouting up and "invading" empty territory. Instead, the colonizer probably fires up preexisting circuits that were lying dormant. Again, the brain has bazillions of neural circuits running every which direction, and some of these tracts happen to start in the face territory and spill over into the adjacent hand region. Most of the chatter on these circuits is irrelevant to the hand, so the hand region mutes them. But when the hand area falls quiet, it loses the ability to resist. The nearby cheek and lip areas suddenly face no opposition and can take over.

Still, as every colonial power in history has learned, occupying a territory is different from assimilating it. Too many "hand circuits" exist to reprogram them all, and the hand territory always retains a vestige of its identity. As a result, the new face circuits and the old hand circuits overlap and intermingle, and both can end up firing simultaneously.

What does this all mean on a higher scale, the scale of perception? It means that for some amputees, touching or moving their faces will summon up sensations in their missing hands. If an amputee strokes his cheek, for instance, he might feel his missing thumb being brushed. If he whistles or chews gum, the index finger twitches. If he pops a chin zit, the pinkie feels the squeeze. Even those people who don't consciously register the dual sensations will still have signals intermixing in the brain. The net result is that face sensations keep stoking the mental memory of the hand and keep stirring the phantom awake.

(Similarly, because the foot and genital territories border each other on the brain map, when the lower leg disappears, the genital spot can take over. Sure enough, some lower-limb amputees feel their phantom feet most insistently during sex. A few even report feeling orgasms quaking all the way down to their phantom tiptoes. And, like striking a bigger tuning fork, this expansion of the orgasmic territory gives them a proportionately greater pleasure.*)

Scientists gained another crucial insight into phantom limbs from a series of almost comically low-tech experiments conducted by a neurologist in Southern California named V. S. Ramachandran. Ramachandran had a patient named D.S. who'd lost his left arm after a motorcycle wreck and had experienced severe phantom cramping ever since. To treat him, Ramachandran took an open-topped cardboard box and mounted a mirror inside. The mirror divided the box's interior into two parts, a left chamber and a right chamber. Ramachandran cut a hole in the box on either side of the mirror and had D.S. slide his right hand into the right chamber. (D.S. also imagined sliding his phantom left hand into the other chamber.) The crucial point is that the reflective surface of the mirror faced right. So when D.S. inserted his hand into the hole and glanced down, it looked like he had two intact hands again.

Ramachandran had D.S. close his eyes and start swinging his hands back and forth symmetrically, like someone conducting the philharmonic. At first nothing happened. The phantom stayed frozen, mute. D.S. then opened his eyes and repeated the motion while looking in the mirror. That's when the orchestra burst into song. As his hands swayed back and forth, his phantom fingers unfurled for the first time in a decade. His cramps abated, his rigid wrists went loosey-goosey. "My god!" he yelled, and began jumping up and down. "My arm is plugged in again."

Over the next few years many more amputees would share that same glee in Ramachandran's office. The "mirror box" looked hokey, sure. But something about *seeing* a lost limb in motion unfroze the phantom in people's minds. Again, we dedicate loads of brainpower to vision, and we implicitly trust sight over our other senses—seeing is believing. So when the eyes see a limb moving again, the brain believes it can.

Based on this and other insights, scientists like Ramachandran have now sketched out an explanation for why phantoms exist and why they often snarl with pain. Because the brain has a hardwired mental scaffold of the body, it expects to find four full limbs at all times—that's its default setting, and that's why even people born without limbs can experience phantoms. Moreover, the reality of phantoms is reinforced when the brain keeps receiving spurious signals, both from the inflamed stump and especially from any greedy brain territories that colonize an empty neural landscape. All this activity fools the brain into thinking that the hand or leg still exists. So the brain keeps sending motor signals down there, and armless men keep grabbing for their hats in gusty weather.

That explains the sensation. The paralysis and pain arise for different reasons. If the limb was paralyzed before the amputation, the phantom is usually paralyzed afterward, too. But even people who can "move" their phantoms at first often lose that ability later. Remember that the brain, after it sends off a movement command, looks for sensory feedback to confirm that movement took place. Arms that don't exist obviously don't provide this feedback. So, over time, most people's brains conclude that the phantom is paralyzed.

Pain can get stamped into a phantom limb just like paralysis does, when preamputation aches and pains carry over. But motor commands can also exacerbate the pain. Because an AWOL limb can't respond to motor commands, the brain—which hates being disobeyed—often ratchets them up: a failed command to *squeeze the left hand* gets transformed into *squeeze hard,* then *squeeze harder,* then *squeeze like hell.* This causes pain for two reasons. One, pain signals alert the body that something is wrong, and with this mismatch between motor commands and sensory feedback, there's clearly something amiss here. Second, brutal commands like those were usually accompanied by pain in the past: your brain learned, for instance, that clenching your fist caused your fingernails to gouge your palm. Eventually the hand-clenching circuit and the pain circuit got wired together. As a result, whenever the brain tries to rouse the phantom with a hard squeeze, pain sensors can't help but fire.

The mirror box, however, slits the neural Gordian knot. It resolves the mismatch between the motor and sensory systems, and because the brain literally sees its commands being obeyed, it can stop sending out orders to *squeeze* and *squeeze harder.* In the sudden tranquility, the pain seeps away. To be sure, at first the relief lasts only a few hours before the phantom seizes up again. Not everyone finds relief from mirror therapy, either. But those who do, and who practice with the mirror box, can see profound improvement over time as their brain maps reorganize. In many cases the pain all but disappears. (You can think about this decoupling as the converse of *neurons that fire together wire together.* Here, *neurons out of sync fail to link.*) And in some cases the phantom itself vanishes. After Ramachandran's first patient, D.S., drilled with the mirror box for several weeks, he felt his phantom left arm shrinking inch by inch, "telescoping" up into his shoulder. Finally, just a nub of sensation remained. Ramachandran called this the first successful amputation of a phantom limb.

~

After publishing his magnum opus on phantom limbs in 1872, Silas Weir Mitchell went on to a career of such brilliance that one admirer declared him "the most versatile American since Benjamin Franklin." He helped pioneer the study of sleep paralysis, traumatic shock, and object blindness. He also resumed his venom research; conducted some, ahem, personal experiments with hallucinogens such as mescaline; and, most infamously, invented the "rest cure" for psychological disturbances, an outgrowth of his interest in helping Civil War vets return to civilian life.

For men, Mitchell's rest cure consisted of a few weeks of roping cattle and sleeping outdoors in the Dakota Badlands or areas farther west. Mitchell prescribed such a retreat, with plenty of mountain air, for his buddy Walt Whitman in 1878, after tracing the poet's dizziness, headaches, and vomiting to a small stroke. Painter Thomas Eakins also underwent this "West cure," and the regimen supposedly cured the young Teddy Roosevelt of his effeminate voice and foppish mannerisms in the 1880s. (Before this, TR was considered soft, and people compared him to Oscar Wilde.) For women, especially for "hysterics," Mitchell prescribed a different sort of rest cure. It consisted of six to twelve weeks of bed rest in a dark room, along with massages, electrical stimulation of the muscles, a sickening surfeit of fatty foods, and complete isolation (no friends, lovers, letters, or novels). As you can imagine, spirited women resented this. After the birth of her daughter and subsequent postpartum funk, Mitchell basically ordered the writer Charlotte Perkins Gilman to stay in bed and stop making trouble: "live as domestic a life as possible," he said, "never touch pen, brush, or pencil again." She responded by writing "The Yellow Wallpaper," a classic feminist story about a woman driven mad by such treatment. (Virginia Woolf gave Mitchell a similar working over in *Mrs. Dalloway*.) Gilman later mailed a copy of her story to

Mitchell and claimed that he amended his ways because of her, but in reality Mitchell continued to condescend to female patients, especially hysterics. When one hysteric refused his orders to end her rest cure, he threatened, "If you are not out of that bed in five minutes, I'll get into it with you." She held out while he removed his coat and vest, but skedaddled when he unfastened his fly. In another case, with a woman faking a mortal illness, he sent all of his assistants out of the room. When he emerged a minute later, he promised she'd be up in no time. How did he know? He'd set her sheets on fire.

In addition to his medical practice, Mitchell began to study medical history, especially the deep and unsettling synergy between war and medicine. As he well knew, only during combat do doctors and surgeons see enough cases of ghastly things like shattered limbs to become experts on them. Moreover, the Civil War prompted great improvements in patient transport, anesthesia, and hospital hygiene. Mitchell's general observation holds for other wars as well. Modern nursing began with Florence Nightingale in the Crimea, and the Franco-Prussian War proved once and for all the importance of vaccines. Later, the Russo-Japanese War sparked important vision research, and World War I improved the treatment of facial injuries. More recently, Korea, Vietnam, and other conflicts taught surgeons how to reconstruct mangled nerves and veins and reattach severed limbs, thus preventing phantoms from arising in the first place. And the recent wars in Iraq and Afghanistan—where close-quarter explosions left thousands of soldiers with low-level but pervasive neuron damage reminiscent of concussions—will no doubt provide their own innovative remedies. However much suffering they produce in the short term, wars have benefitted medicine profoundly.

Even as his scholarly and scientific reputations were peaking, Mitchell felt more and more drawn to another pursuit—writing. His clinical papers on nervous ailments had always felt dehumanizing: too liable, in their pursuit of general truths, to trample an individual's story. In contrast, fiction writing let Mitchell capture the nuances of a man's life, and capture the way he experienced something like phantom limbs. Mitchell was actually taking part in a broader literary movement: Balzac, Flaubert, and others also poached on medical work to heighten realism and draw more convincing portraits of suffering. Nevertheless, fiction writing wasn't deemed a respectable hobby for physicians in those days, and Mitchell's friend (and fellow doctor and writer) Oliver Wendell Holmes Sr. advised him to keep his writing on the down low, since patients wouldn't trust a doctor who used them as fodder.

Only in the 1880s, after twenty years of publishing anonymously, did Mitchell come out of the authorial closet. Thereafter his scientific work tapered off, and he began writing almost full-time, eventually publishing two dozen novels. He often saddled his characters with seizures, hysteria, split personalities, and other nervous ailments. And although he wasn't above tossing in a ghost to enliven the plot, he wrote mostly realistic works with an emphasis on moral dilemmas. Teddy Roosevelt declared Mitchell's bestseller *Hugh Wynne: Free Quaker* probably the most interesting novel he'd ever picked up. And toward the end of his life, at age seventy-five, Mitchell finally owned up to writing "The Case of George Dedlow" four decades earlier. Mitchell had taken Dedlow's name from a jeweler's shop in a Philadelphia suburb, mostly because he found it apt ("dead-low") for a double leg amputee. He'd sent the story to a female friend for feedback. Her father, a doctor, read about phantom limbs with fascination, and forwarded it to the editor of *The Atlantic Monthly*. Mitchell claims he forgot about the story until the page proofs and an $85 check arrived in the mail. Regardless, the story's success galvanized him. At that point he hadn't published anything academic about phantom limbs, and without the public outpouring for Dedlow he might never have pushed his fellow doctors to take phantom limbs seriously.*

A friend once noted of Mitchell that "every drop of ink [he wrote] is tinctured with the blood of the Civil War." Even on his deathbed—in January 1914, as the world prepared for a new war in Europe—Mitchell's mind could only drift back to Gettysburg and Turner's Lane. He in fact spent his last, delirious moments on earth conversing with imaginary soldiers in blue and gray, pursuing phantoms to the end.

CHAPTER SIX

The Laughing Disease

So far we've mostly considered one-way communication—from brain to body, for example. But the nervous system also uses feedback loops, to tweak commands on the fly and combine signals in new and sophisticated ways.

Toward the end the victims laughed frantically, explosively, on the slimmest pretense, laughed so hard they'd fall over and sometimes almost roll into the fire. Until that point their symptoms—lethargy, headaches, joint pain—might have been anything. Even when they began to stumble about and had to flail their arms in a herky-jerky dance to stay balanced, even those tics might be explained away as sorcery. But laughing could only mean kuru. Within months of the first symptoms, most kuru victims—predominately women and children in eastern Papua New Guinea—couldn't stand upright without clutching a bamboo cane or stake. Soon they couldn't sit up on their own. When terminal, they'd lose sphincter control and the ability to swallow. And along the way, many would start to laugh—laughing reflexively, senselessly, with no mirth, no joy. The lucky ones died of pneumonia before they starved. The unlucky ones were whittled down until their ribs pushed through their skin, and the women's breasts hung deflated.

After a few days of mourning, the local women raised the victim on a stretcher of sticks and bark, and gathered in a secluded bamboo or coconut grove distant from the men. Silently, they started a fire and greased themselves with pig fat to protect against the insects and the nighttime cold of the mountain highlands of Papua New Guinea. They laid the body on banana leaves and began sawing each joint, fraying the cartilage with rock knives. Next they flayed the torso. Out came the clotted heart, the dense kidneys, the curlicue intestines. Each organ was piled onto leaves, then diced, salted, sprinkled with ginger, and stuffed into bamboo tubes. The women even charred the bones into powder and stuffed that into tubes; only the bitter gall-bladder was tossed aside. To prepare the head they burned the hair off, gritting through the acrid smell, then hacked a hole into the skull vault. Someone wrapped her hands in fern leaves and scooped out the brains and filled still more bamboo. Their mouths watered as they steam-cooked the tubes over warm stones in a shallow pit, a cannibalistic clambake. In dividing up the flesh, the victim's adult relatives—daughters, sisters, nieces—claimed the choicer bits like the genitals, buttocks, and brain. Otherwise, people shared most everything, even letting their toddlers partake in the feast. And once they started feasting, they kept stuffing and stuffing themselves until their bellies ached, taking leftovers home so they could binge again later.

The tribe never named itself, but explorers called them the Fore *(For-ay),* after their language. In Fore theology, consuming someone's body allowed his or her five souls to enter paradise more quickly. Moreover, incorporating their loved ones' flesh into their own flesh comforted the Fore, and they considered this more humane than letting maggots or worms disgrace someone. Anthropologists noted another, more prosaic reason for the feasts. For food, the Fore mostly gathered fruits and vegetables and scraped a few *kaukau* (sweet potatoes) out of the poor, thin mountain soil. A few villages kept pigs, and hunters speared rats, possums, and birds, but the men usually hoarded these spoils. The funeral feasts let women and children gorge on protein, too, and they especially enjoyed eating kuru victims. Kuru left people sedentary, unable to walk or work, and those who died of pneumonia (or were euthanized by smothering before they'd starved) often had layers of fat.

Despite the feasts, kuru—from a local word for "cold trembling"—alarmed the Fore, and they concealed its existence from the outside world for decades. Doing so wasn't hard, as they lived in the eastern highlands of New Guinea, among the most isolated places on earth; through the mid-1900s, many tribes there didn't know salt water existed. But soon enough the outside world began wrapping its coils around the Fore and other nearby groups. Gold miners tramped through the highlands in the 1930s, and a Japanese plane crashed there during World War II. Missionaries dribbled in, and in 1951 Australia established a patrol post for men who enjoyed wearing short khaki shorts and pointing rifles at people who lacked even metal tools. Kuru had reached epidemic levels by then, but most of the outsiders were worried about other things, like the tribes' excessive violence and their outré sex habits. (One-quarter of adult males in the highlands died in raids or ambushes, and some tribes initiated boys into manhood with ritual sodomy.) Some white visitors did catch a glimpse of a kuru invalid being hustled out of sight now and then, or noticed the curious lack of burial grounds in a place with such high mortality. But even the first western doctor to examine a kuru patient arrived at the rather Victorian diagnosis of hysteria, hysteria fueled by colonialism and the erosion of traditional tribal life.

The more cases of kuru that emerged, though, the more empty that diagnosis seemed. How could a seven-year-old with no memories of tribal life come down with hysteria, much less die of it? Kuru was clearly organic, and the movement and balance problems suggested brain trouble. But whether kuru was genetic or infectious, no one knew. To compound the mystery, unlike all other known infections or neurodegenerative diseases, which don't discriminate by race or creed, kuru attacked only the Fore and their neighbors, some 40,000 people; *The Guinness Book of World Records* once named kuru the rarest disease on earth. But exactly because of its oddities, this rarest of diseases soon became a global obsession, with samples of Fore brains speeding across the globe and opening up whole new realms of neuroscience.

~

The highlands attracted a strange breed of visitor. People who laughed off leeches and lice. People who didn't mind that the natives greeted them by fondling their breasts or sprinkling pig's blood on them. People who shrugged when the roads were washed out yet again and didn't blink when told that reaching a village a few miles distant would require an eight-hour hike around gorges and cliffs. You almost had to thrive on hardship, and throughout the 1950s New Guinea attracted its share of misfits—none more misfit than D. Carleton Gajdusek.

Born to a butcher in New York State, Gajdusek *(GUY-duh-sheck)* proved a science prodigy as a boy. He sailed through school, and on the stairs leading up to his lab in the attic he painted the names of Jenner, Lister, Ehrlich, and other great biologists. (A dubious legend had it that he left the top stair blank, for himself.) Still, he had trouble relating to his peers, to say the least; he once threatened to poison his entire class with the cyanide his aunt had given him to collect bugs. So at age nineteen this young man with icy blue eyes and pitcher ears ran off to Harvard Medical School, where he earned the nickname Atom Bomb for his intensity. He specialized in pediatrics, then did graduate work in California on microbes. His circle of colleagues there included James Watson.

Neuroscientist and adventurer Carleton Gajdusek. (National Library of Medicine)

But just as Gajdusek began to establish himself in American science, he began to chafe at the conventions of bourgeois American life. He finally escaped under the auspices of the army medical corps and set about wandering through Mexico, Singapore, Peru, Afghanistan, Korea, Turkey, Iran. At each stop he hunted out children with rabies or plagues or hemorrhagic fevers, doing pathbreaking work on little-known diseases. He made friends easily and lost them even more easily, often in blowout fights. In fact he had little personal life beyond his pediatric work: a colleague once observed that he had "no interest in women, but an almost obsessional interest in children." Like the Pied Piper, he attracted a coterie of boys in every remote village, and he once wrote in his journal, "Oh, that we might be Peter Pans and live always in Never-Never Land."

In early 1957 he visited New Guinea, planning to cruise right through—until he heard about kuru. Kuru combined his interests in microbiology, neurology, children, and remote cultures, and the colleague who first briefed Gajdusek about it compared his reaction "to showing a red flag to a bull." Gajdusek caught the next bush plane up to the highlands and began tramping from village to village over some of the steepest, slipperiest terrain on earth. He quickly memorized the symptoms—twitching eyes, a staggering gait, trouble swallowing, laughter—and identified two dozen kuru victims within a week, sixty within a month. With growing excitement he also began writing letters to colleagues, alerting them to this new disease.

He spent the next few months conducting a kuru census, visiting every village he could and taking tissue samples from victims. To this end he recruited—with soccer balls and other toys—an entourage of ten- to thirteen-year-old *dokta bois* (doctor boys), dozens of whom might accompany him on a patrol. They marched for hours with Gajdusek every day, clad in white *laplaps* (waistcloth skirts) and carrying boxes of rice, tinned meats, and medical supplies on poles over their shoulders. They had to dodge bees and mudslides and stinging plants. They made tea in streams and wielded bamboo torches after dark.

Their shelters for the night were often barely distinguishable from the surrounding shrubbery, and they lived in perpetual fear of ambushes from neighbors with bows and arrows. Reaching some villages required crossing gorges on bamboo bridges that disintegrated with each step, the chaff flaking off and floating down a hundred feet to the rivers below. Naturally, most boys saw the patrols as grand adventures, the happiest hours of their lives.

At every stop Gajdusek asked about kuru, and the more enterprising *dokta bois* snuck into the bush to rustle out victims who'd been hidden away. Some boys were beaten for this by family members who wanted their mothers and aunts and children to die in peace. But whenever a victim agreed, Gajdusek took blood and urine samples in makeshift bamboo tubes and packed them away in his supply boxes.

After a thousand miles of hiking, Gajdusek had determined just how bad things were. Roughly 200 people were dying of kuru annually, the proportional equivalent of 1.5 million U.S. deaths every year. And things were actually worse than that sounds. Because kuru

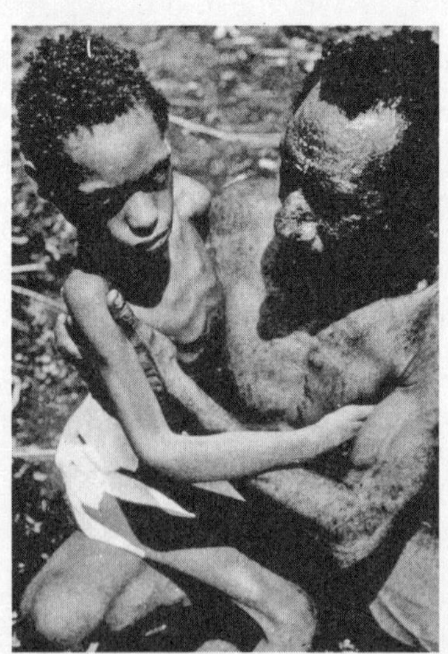

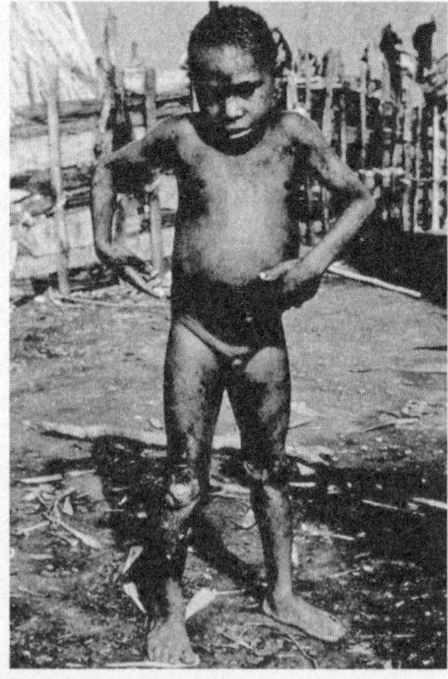

Two young kuru victims. (Carleton Gajdusek, from "Early Images of Kuru and the People of Okapa," *Philosophical Transactions of the Royal Society B* 363, no. 1510 [2008]: 3636–43)

targeted women and children, it threatened to extinguish the Fore culture, since the younger generation couldn't replenish itself. More acutely, the chronic shortage of women, a common cause of war among hunter-gatherers, seemed likely to ratchet up tensions even more.

The delicacy of the situation made the ruling Australian government tremble. Australia had acquired the highlands after World War I, and politicians there viewed New Guinea as their one chance to become a colonial power. As with most colonial overlords, Australia was motivated by a patronizing wish to "civilize" the natives, combined with a strong lust for profit, and by 1957 it had achieved both ends. Fewer and fewer natives wore penis sheaths or pierced their noses with pig tusks. Papuans now built rectangular homes instead of traditional oval ones, and they abandoned their simple, bamboo-pipe-irrigated yam gardens to slave away on coffee plantations or in mines. At the same time murder rates had dropped off steeply and centuries-old diseases like yaws and leprosy had disappeared. But kuru threatened to upset this *pax Australiana* by panicking the highlanders and discrediting the government. Colonial officials tried to keep it secret, and they despised Gajdusek for spreading word of it. Hell, for all they knew, Gajdusek himself was spreading the disease by tramping from village to village. So colonial officials tried to restrict his movements within the highlands and even petitioned the U.S. State Department to forbid his travel. They meanwhile played dirty and waged a propaganda war, denouncing him as a "scientific pirate" and threatening other scientists for collaborating with him. One rival taunted Gajdusek that "your name is [now] mud."

But Australia was about to learn that Carleton Gajdusek did not lose stare-downs. After throwing a tantrum over the interference, he decided to simply outwork his saboteurs. He'd penetrate more deeply into Fore territory and collect more gallons of blood, urine, and saliva than any five Aussies. Sure enough, within five months Gajdusek had identified hundreds of kuru victims, and he even sweet-talked some families—or bribed them, with knives, blankets, salt, soap, and tobacco—into letting him do autopsies on the victims' brains. Like a pseudo-cannibal himself, Gajdusek performed some of these autopsies on the kitchen table in his hut, plopping the brains onto his dinner plates and slicing them up like thick white focaccia with a gray-matter crust. He sent most of this precious tissue back to his lab at the National Institutes of Health, in Maryland, but shrewdly also sent samples to Australian scientists, to placate them and undermine the politicians whispering poison into their ears. Eventually Australia realized it would just have to tolerate Gajdusek.

In the meantime Gajdusek faced another, unexpected obstacle to his work—sorcery. Almost to the last man and woman, the Fore believed that sorcerers caused kuru, and they listened to Gajdusek's lectures on microbes and genetics with amuse- or bemusement. According to tradition, sorcerers worked their necromancy on personal items, including body discards such as hair, fingernails, and feces. Sorcerers first bound these items with leaves, then cast their spells and buried the bundles in swamps; as the items decayed, so too would the victims' health. (To be sure, the Fore considered most spells cast in this manner perfectly acceptable, but "making kuru" went beyond the bounds of decency.) To head off sorcerers, the Fore held bonfires to burn their refuse, and also built some of the deepest latrines on the planet. (After doing their business in the woods, they might even carry the turds back to the latrine, for safety's sake.) And people who had already contracted kuru would hire showy counter-sorcerers, who chanted and dispensed herbs and forbade patients from drinking water, eating salt, or consorting with the opposite sex. Not surprisingly, people who believed so deeply in sorcery weren't thrilled about giving up bodily fluids to a stranger. So to convince people of their security, Gajdusek acquired a reassuringly large lock, which he slapped onto his box of samples.

After Gajdusek collected the samples, they had a dicey future. If he had access to a jeep, he drove them to the nearest patrol station.

Often as not, though, an axle had broken or the road had washed out, and he had to dispatch a *dokta boi* on a multihour hike. It was then even odds whether the freezer at the station was working. Within a few days, hopefully, the blood or brains got loaded onto a plane en route to a city with an international airport. There, a technician could finally pack the samples in dry ice and dispatch them to Maryland or Melbourne or the dozen other places where labs—goaded by Gajdusek—had taken up kuru.

Neurologists also started trickling into the highlands to examine kuru victims directly and look for signs of brain damage. Some of the tests they administered looked like DUI checkpoint trials, with Fore people walking heel to toe, touching fingers to noses, or standing on one leg and raising both arms. Kuru victims generally failed such tests. Neurologists also tested for certain reflexes. If you tap the skin around an infant's mouth, she'll automatically purse her lips; this "snout reflex" makes it easier to suckle a nipple. Similarly, brushing an infant's palm in certain places will make her fingers curl, a reaction called the hand-grasp reflex. These reflexes disappear during our second or third year as the brain matures and certain other circuits inhibit them. But they can reemerge after brain damage—and often did in kuru victims.

Based on the battery of tests, neurologists traced much of the initial damage in kuru to the brain's movement centers, especially the cerebellum. As we've seen, a few different patches of gray matter in the brain (e.g., the motor cortex) work together to initiate movement. In addition, the brain's motor system has some crucial feedback loops to ensure that movements are carried out properly. One key structure in this feedback circuit is the cerebellum.

As part of the so-called reptile brain, the cerebellum sits way back near the spinal cord, and its wrinkly appearance makes it look like a mini-brain all by itself.* It plays an especially important role in coordinating movement and providing balance. In short, the cerebellum collects inputs from all over the brain, including all four lobes. This allows it to monitor the body's position in space in multiple ways (through touch, vision, balance, and so on). It then checks to see whether the movement you're executing is anything close to what you intended. If not, the cerebellum pings another brain structure (the thalamus), which passes the message to the motor cortex and tells your muscles how to adjust. *Not so fast,* it might caution, or *a tad to the left.* Without the cerebellum you might get lucky and grab your glass of wine every so often, but it's more likely that you'd flail your arm too far one way, then correct wildly the other way and knock the glass over. In other words, the cerebellum makes grace and precision possible. It helps control the timing of movements, enabling you to walk, talk, jump, and swallow smoothly. Even some involuntary movements, like breathing, depend on the cerebellum to some extent.

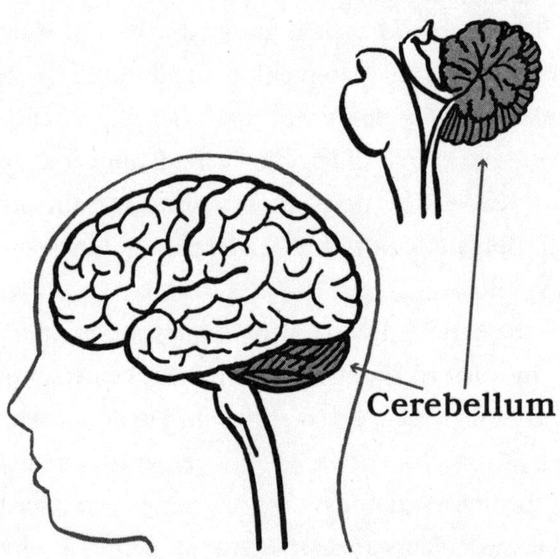

When the cerebellum deteriorates, then, your balance falters and your movements become clumsy. Hence the trembling, the eye-twitching, and the jerking gait of kuru victims. Pathological laughter can also arise when a circuit that involves the cerebellum suffers damage. And of course degenerative brain diseases rarely confine themselves to one spot in the brain. The snout and grasp reflexes, and the general cognitive decline of victims, told neurologists that kuru eventually radiated outward and affected structures like the frontal lobes.

Even as the anatomical damage became clear, though, the underlying cause of kuru remained murky, especially on a molecular level. Some scientists jumped to the conclusion that because kuru often ran in families, it must be genetic. But as Gajdusek knew, this theory had holes. For one thing, kuru spread not only within families but also sometimes from one unrelated adult to another, which isn't genetic behavior. Moreover, adult males almost never got kuru, while adult women did frequently. That might suggest something sex-linked—except that the incidence was equal among prepubescent boys and girls. Gajdusek suspected that kuru spread via infection. But that theory ran aground on the fact that the brains he autopsied showed no inflammation and zero other signs of infection.

Nevertheless, the autopsies did reveal other clues. In 1957 an American colleague of Gajdusek's discovered "plaques" in the brains of kuru victims—gnarly black burrs of protein a thousandth of an inch in diameter. The colleague also noticed a proliferation of astrocytes, a type of glial cell shaped like a star. Close to half the cells in the brain are astrocytes, and they play an important role in forming the blood-brain barrier, a protective sheath around blood vessels that blocks foreign material from entering the brain. But for whatever reason, astrocytes also multiply beyond control in the gray matter whenever neurons die off, eventually forming scars. The colleague had no idea what might be causing the protein plaques and astrocyte scars in kuru victims, but he did note a resemblance to Creutzfeldt-Jakob disease (a.k.a. human "mad cow" disease).

Two years later came another clue, from across the Atlantic Ocean. After a friend recommended he visit, an American veterinarian named William Hadlow attended a museum exhibit on kuru in London. He wandered about among the Fore artifacts, interested but hardly engrossed, until some photographic blowups of kuru brains arrested him. The tissue in the pictures looked oddly spongy, and oddly familiar. Hadlow had studied scrapie, a disease that wrecks the brains (especially the cerebellums) of sheep, causing them to stagger and to scrape their skin raw on trees or fences. Some sheep even hop like rabbits. Scrapie-infested neurons have holes in them, as if tiny carnivorous moths had invaded. Scrapie brains also contain larger holes, where whole bundles of neurons have died. Hadlow noticed that kuru brains had the exact same patterns of holes—the exact same spongy appearance. He quickly wrote a paper, and Gajdusek got in touch with him shortly thereafter. As with the Creutzfeldt-Jakob link, the scrapie connection was a great lead but frustrating, since no one knew what—toxins? genes? viruses? some combination?—caused either disease.

~

The words "slow down" had no meaning for Gajdusek, but with so many other scientists now studying kuru, he did decide to indulge some of his other interests, especially anthropology. He built a bamboo hut for himself in the eastern highlands and began documenting life there, taking thousands of photographs and literal miles of reel-to-reel video. Despite the constant mist and paper-curling humidity, he also filled 100,000 journal pages with field notes about most everything under the sun—local songs, etymology, bawdy gossip, recipes, the inroads made among the locals by communism and Christianity. He used the journals as a diary as well, recording how much weight he'd lost in the field (25 pounds, down to 160) and his fantasy that he could see *Sputnik* circling among the stars overhead.

As a child-development specialist, one subject above all obsessed Gajdusek—sexual initiation rites—and he traveled widely across the highlands, far beyond the Fore tribes, to gather information on them. At around age seven, boys in some highland tribes moved into special domed huts, where they spent their days and nights servicing older adolescents and occasionally grown men. "You should not be afraid of eating penises," the elders instructed them: they believed that semen strengthened young boys the way breast milk strengthened infants. Oral and anal sex also allowed the boys to "store up" semen, since some tribes believed that men didn't produce semen naturally. Gajdusek recorded every rite he could, down to the details of who got frisky with whom. He also marveled at how some boys even "flirted" with him, batting their eyes and stroking his pasty skin. In his field notes he emphasized that the tribes sanctioned all this underage sex, and argued that it served an important social function in keeping the men from warring over women. (Other anthropologists rolled their eyes at such interpretations.) What's more, the rites helped Gajdusek realize that the "stuffy" sexual mores of the world he'd grown up in were not universal.

In fact the more he immersed himself in highland culture, the more Gajdusek found his past life wanting. He never renounced Western civilization entirely: in particular, he devoured the decadent literature of Henry James and Marcel Proust during his spare hours in the field. But right in the middle of some passage about dukes and duchesses, he might glance up to see Papuan youths dancing outside his hut, wearing feathered headdresses and tusks through their noses. Like Gauguin, this primitive life attracted him, and the dueling impulses—the intellectual and the primal—warred for his soul. One colleague remembered him disappearing into the bush for weeks, then stumbling into a dinner party after his journey in a dirty T-shirt and shorts, one shoe missing. However disheveled, he always dazzled guests with repartee—bouncing until 4 a.m. from Melville to meadow mice to Plato to Puritanism to suicide to Soviet foreign policy—all before disappearing into the wilderness again. Like Kurtz in *Heart of Darkness,* he seemed to be wrestling with all of Western civilization.

Meanwhile, the Fore had their own bone to pick with Western civilization, particularly with Western medicine. Doctors had recently used "shoots"—injections of medicine—to eradicate leprosy in the area. Although thankful, the natives didn't take this as a sign of the superiority of Western science; rather, they concluded that Western doctors must be powerful sorcerers indeed, far more powerful than the native sorcerers who caused diseases. So when doctors set about trying to cure kuru, Fore expectations were high. Unfortunately, none of the vitamins, tranquilizers, steroids, antibiotics, liver extracts, or other medicines that Gajdusek and company hauled into the field did any good: kuru always killed. After years of futile interventions, the Fore began to seethe. The white men took, took, took, they complained—took bodies, took blood, took brains—but gave nothing back. Even those who believed in Western medicine lashed out. One of Gajdusek's companions fumed that he knew America had "the big microscope" capable of curing any disease, and he couldn't understand why Gajdusek didn't hurry the hell up with kuru.

As the situation unraveled, the Australian government, desperate to stop kuru, considered building a giant fence around the Fore and confining them to a "reserve." (Not only would the fence keep the Fore in, they noted, it would keep Gajdusek out.) Inspired by the genetic theory of kuru, officials also discussed sterilizing the tribe.

But with each new victim, it became clearer that the genetic theory didn't hold water: kuru simply spread too quickly, killing most people before they'd passed on their genes. Plus, some women who were genetically distinct from the Fore and had only married into the tribe had also come down with kuru.

At the same time, no other possible cause made sense. Kuru was neurological, clearly. But scientists had failed to find any bacteria or viruses in the brains of victims. Other experiments ruled out hormone imbalances, autoimmune diseases, metal toxins, plant toxins, insect toxins, alcoholism, and STDs. Some doctors suggested cannibalism as a factor, but the practice had already been outlawed by then. Besides, the Fore had always cooked the bodies thoroughly before eating them, and their customs forbade children from dining on brains anyway, because eating brains supposedly stunted their growth.

With the Fore growing testier and testier, doctors in the field resorted to bartering treasure for tissue, which made for some ugly scenes. Often the doctors would camp outside a village that had a terminal kuru case, erecting a few poles and throwing up a tarp for a makeshift autopsy clinic. At the first howl of mourning they'd enter the family's hut and start negotiating, offering axes, blankets, tobacco, salt cakes, even American cash. One man argued that if white men took his "meat" (his wife's brain), he should get meat in return. The doctors rustled up a three-pound ham—at which point the husband thanked them, joined the mourners outside, and wailed louder than anyone. The autopsy often took place under kerosene lamps or in drizzling rain, and it could take hours of slicing, cracking, and sawing to liberate the brain and spinal cord—an eon in a place with spotty refrigeration. Doctors wrapped the autopsy up by stuffing cotton balls into the skull and returning the body. They then had the distasteful job of making sure the villagers buried the body instead of eating it.

As for Gajdusek, he continued his anthropologic-cum-medical work, and despite admonishing himself not to, he found himself more and more entangled with the personal lives of his patients. One sad incident involved Kageinaro, a young boy. Although frisky and "flirty" on previous encounters, Gajdusek entered Kageinaro's village one day to find him acting aloof and distant. Gajdusek asked the boy's friend what the matter was. The friend sighed, "Me tink 'e gat sik." *Sick.* "All at once," Gajdusek recalled, "I knew another of my boys had kuru." That night he insisted Kageinaro sleep beside him, for comfort; he wrote in his journal the next morning that "if kuru is contagious I certainly have it." Gajdusek also returned months later to be with Kageinaro as he died, pulling him out of the soiled "lair" where his family had abandoned him. The boy reeked; his eyes flinched from the sunlight; he turned his crusted mouth away from Gajdusek in embarrassment. Gajdusek comforted him as best he could, holding him and giving him water. Most of it ran in rivulets down Kageinaro's cheeks, since he couldn't swallow. Gajdusek, mirroring the boy's face, sobbed.

Scientists soon entered Kageinaro's name in The Book. This pile of white looseleaf sheets, bound together and carried in an attaché case, acted as a sort of Papuan Doomsday Book, recording every known kuru victim from 1957 forward. As a scientific document The Book is a marvel—scientists had never tracked a disease with such precision. As a social document The Book is simply sad, an unparalleled chronicle of devastation. It records that 145 of the 172 hamlets in the area lost someone to kuru, and some villages lost 10 percent of their women in one year. Reading between the lines, the entire social order was crumbling, and while the *dokta bois* worked and worked, hauling the brains of loved ones to patrol stations and even visiting the villages of enemies to collect samples, The Book just kept getting fatter and fatter. It eventually swelled multiple inches thick.

A breakthrough came at last in the mid-1960s. Although focused on fieldwork, Gajdusek maintained an active research lab back in Maryland. Tantalized by the possible connections among the diseases, he and his team of scientists started to inject cells infected with kuru, scrapie, and Creutzfeldt-Jakob disease into the brains of rodents, to determine whether those diseases were contagious. (Getting scrapie into the United States had required him to flout an international ban and smuggle the tissues in personally, but Gajdusek never considered himself bound by petty laws.) These diseases were indeed contagious, so in 1963 he took the next step, gathering a cohort of apes in an uninsulated cinder-block building in rural Maryland.

Not long before, a boy named Eiro and a girl named Kigea had died of kuru in Papua New Guinea. Near the end both could do little but grunt, and they'd subsisted for weeks on sugar water. (When Kigea's doctor offered her a lollipop, she was too weak to grasp it.) Their families agreed to autopsies, and thanks to a wondrous new material, Styrofoam, their nice, cold brains reached Maryland in pristine condition. On August 17, 1963, Gajdusek and colleagues mixed a thousandth of an ounce of Kigea's brain with water and injected the slurry into the skull of a chimp named Daisey. Georgette the chimp got an injection of Eiro's brain matter four days later.

As they settled in to monitor the chimps' health, the team had to fend off the U.S. Department of Agriculture, which wanted to know why the hell someone was messing around with biological agents in an unsecure building in Maryland farm country. Meanwhile, Gajdusek, never one to sit idle, kept traveling around the world and directing various other research projects from his anarchic NIH laboratory. Visitors remember Bob Dylan blasting on the stereo, psychedelic posters brightening the walls, and lab assistants practicing yoga.

To maintain his connection to New Guinea in between trips there, Gajdusek also began "adopting" Papuan youths, starting with a brash and lively lad named Mbaginta'o in 1963. Among other things, Mbaginta'o had to learn how to use a toilet, wear shoes, and eat with utensils before immigrating to Maryland. Gajdusek nevertheless enrolled him, as "Ivan" Gajdusek, in Georgetown Prep, an elite local high school. Ivan adjusted well, and eventually Gajdusek brought over a "brother" for him. He too thrived, so another brother appeared. Then another. Pretty soon—story of his life—Gajdusek went overboard, and dozens of teens from other tribes followed over the next few decades, some initially against their will. Gajdusek paid to feed and clothe all of them and sent them to good schools. Rather than focus on academics, however, many of his "sons" preferred boozing, racing cars, seducing the daughters of local Rotarians, and generally, as Gajdusek fumed, "fucking around." In short, they behaved like teenagers. Gajdusek did invoke some discipline—his boys did laundry, mowed, cooked, and cleaned their rooms. But it didn't help them adjust when "dad" would jet off for months at a time to track down some exotic disease and occasionally leave them unsupervised.

The breakthrough came in 1966. After years of tedium—and no results—Daisey the chimp developed a drooping lip and a shuffling, stuttering gait, signs of cerebellum damage. Georgette came down with symptoms shortly afterward. After drawing blood and ruling out every disease and nutritional deficiency and poison they could think of, colleagues summoned Gajdusek back from Guam. Gajdusek arrived grumpy—he hated having his trips cut short—but grew excited when he saw the chimps. The researchers euthanized them and performed autopsies, then sent some brain tissue to a pathologist. She found plaques and spongy holes. Gajdusek's team ripped off a paper for *Nature* in one day, and it appeared in print two weeks later, exploding like a grenade. Not only had they killed the genetic theory of kuru, they'd proved that a degenerative brain disease was contagious in primates, an unheard-of result. Furthermore, they dared speculate on the broader implications of their work for medicine. They proposed that kuru, scrapie, and Creutzfeldt-Jakob—which all cause "spongiform" brain damage and can all lie dormant for long periods before roaring awake—were caused by a new class of microbes, which they dubbed "slow viruses."

The epidemiology of kuru also became clearer in the 1960s. Gajdusek had always balked at linking kuru and cannibalism, since doing so reinforced barbaric "bushman" stereotypes.* Besides, the cannibalism connection had always foundered on a few facts. For one, only women ate the brains at funeral feasts, but children still got kuru, children of both sexes. What's more, Christian missionaries—despite insisting that the Fore eat the body and blood of Christ—had all but eradicated cannibalism by the mid-1950s, while kuru itself had not ceased.

To some people, though, cannibalism still made a lot of sense. The Fore had adopted cannibalism only in the 1890s, when the fad of funeral feasts had diffused down from the north. Intriguingly, the first kuru cases appeared a decade later. And kuru flared up hottest among the tribes most enthusiastic about the feasts. More important, dogged anthropologists determined that the Fore had been lying a little about who ate what. Gooey gray and white matter were supposed to be verboten for the young'uns, but Fore mothers, being mothers, had often indulged them anyway, providing a plausible vector for the infection. And while cannibalism did cease in the 1950s, the chimp experiments explained the lag, since kuru could take years to emerge even when injected directly into the brain. With all these facts laid in front of them, scientists realized that cannibalism explained everything.

These were the first bits of good news, ever, in kuru research. Thankfully, they weren't the last. By the late 1960s the demographics of kuru had shifted and it was becoming rarer. With no more funeral feasts, the average age of victims in The Book was increasing year by year, as fewer and fewer young people contracted it. Kuru never quite disappeared, but by 1975, when Papua New Guinea gained independence from Australia, the highlanders could finally feel they were putting an awful three-quarters of a century behind them.

What's more, in 1976, their champion, Gajdusek, won a Nobel Prize for his discovery of slow viruses. Gajdusek led an American sweep in the sciences that year, and Milton Friedman and Saul Bellow won as well. Gajdusek got characteristically pissy about all the fuss and formality of the prize. (Friends speculated he'd probably never worn a tie before the ceremony.) But the Nobel affirmed kuru as a disease of major importance. Besides, Gajdusek got a kick out of taking eight of his adopted boys to Sweden. They slept in one of Stockholm's fanciest hotels—on the floor, in sleeping bags.

~

Even with the imprimatur of a Nobel, however, one question kept nagging scientists: what exactly were the slow viruses that caused kuru, scrapie, and Creutzfeldt-Jakob?

One problem with the slow virus theory was the presence of the blood-brain barrier. Scientists have known since 1885 that if you inject, say, blue dye into the bloodstream, the heart, lungs, liver, and pretty much every other organ will turn blue. The brain won't, because the BBB allows only certain preapproved molecules across. (Unfortunately, it also bars most pharmaceuticals we swallow or inject, making common brain diseases like Alzheimer's and Parkinson's difficult to treat.) Microbes have an even harder time crossing the barrier: aside from some exceptions, like the corkscrew-shaped syphilis bacteria that plagued Charles Guiteau, most bugs cannot penetrate the neurosanctum neurosanctorum.

Furthermore, kuru brains never got inflamed, a fact impossible to square with any known microbe. The purported viruses proved alarmingly resistant to sterilization, too. Tissue infected with kuru remained contagious even after it was roasted in ovens, soaked in caustic chemicals, fried with UV light, dehydrated like jerky, or exposed to nuclear radiation. No living thing could survive such abuse. This led a few scientists to suggest that the infectious agents might not technically be alive; perhaps they were mere scraps of life, like rogue proteins. But this idea ran so counter to everything biologists knew that getting them even to consider it took someone every bit as tenacious and stubborn as Carleton Gajdusek.

That someone was Stanley Prusiner, who initiated the next great phase of kuru research. Not that his career had gotten off to a great start. Prusiner, a neurologist, pretty much flunked out when he first visited the highlands in 1978. Native *bois* practically had to push him up the mountains with two hands on his backside, and not long after seeing his first patients, some intestinal distress waylaid him and villagers had to lug him back down. Nevertheless, Prusiner returned to his lab in San Francisco full of grand plans. In particular he bet big on rogue proteins as the biological vector for both kuru and Creutzfeldt-Jakob. Unlike cells, proteins aren't alive; in fact, most proteins are helpless outside the cell. But maybe, just maybe, Prusiner argued, some proteins could survive independently and even reproduce in some manner. Because they're simpler, proteins should also survive sterilization better, should have an easier time crossing the BBB, and should avoid triggering inflammation in the brain, since they lack the proper markers for our immune cells to recognize.

Somewhat rashly, Prusiner decided—even before he had any evidence that they existed—to name these rogue proteins, calling them prions *(pree-ons),* a portmanteau of *pro*tein and *in*fection. (This fudged the order of the *i* and *o,* of course, but Prusiner felt the ends justified the spelling peccadillo. "It's a terrific word," he once gushed. "It's snappy." Certainly snappier than *proins.*)

Most scientists disparaged the prion as a vague, fictitious construct—the "p-word," they called it. And in parallel with their dislike of prions, many colleagues developed a pretty healthy aversion to Prusiner himself. In some circles the p-word came to stand for *pushy* and *publicity,* since Prusiner preened and promoted himself and even hired a PR agent. To be fair, Prusiner repeatedly offered to collaborate with colleagues, but most spurned him, including Gajdusek's group. Another time, when Prusiner named Gajdusek as a coauthor on a paper, as a courtesy, Gajdusek hijacked the writing process and refused to allow Prusiner to publish it until he'd deleted all mention of the word "prion." To his credit, Prusiner shrugged these insults off. And after years of laborious work, his team finally isolated a prion in 1982.

The discovery almost ruined him. During follow-up work, his lab determined that normal brain cells manufactured a protein with the exact same amino acid sequence as the prion protein. (Amino acids are the building blocks of proteins.) In other words the healthy brain, as a matter of course, produced something pretty much identical to prions all the time. But if that was true, why didn't we all have kuru or Creutzfeldt-Jakob? Prusiner didn't know, and he brooded over this reversal for months.

Never one to get too discouraged, he soon realized that, far from invalidating his theory of infectious proteins, this new result made it all the more interesting. The key point is that, while the amino acid sequence does help define a protein's identity, proteins are also defined by their 3-D shape. And just as you can rearrange the same sequence of fifty Legos into different structures by snapping the pieces together at different angles, the same sequence of amino acids can be twisted into different proteins with different shapes and different properties. In this case, Prusiner's team determined that a crucial corkscrew-shaped stretch on the normal prions—the ones healthy cells made—got mangled and refolded in the deadly prions, like an untwisted coat hanger. Clearly, there was a "good" prion and a "bad" prion, and kuru and Creutzfeldt-Jakob seemed to involve the conversion of the former to the latter.

So what causes the conversion? Oddly, the catalyst turns out to be the bad prion itself. That is, the bad prion has the ability to lock onto copies of the normal prion that float by and mangle them, changing their shape until they're clones of the bad one. These bad clones then clump together, forming minuscule protein plaques that harm neurons. That's bad enough, but every so often the clump grows too large and breaks in two. And when it does—here's the key—the rate of converting good prions into bad prions doubles, since each half can now drift off and corrupt others independently. Even worse, those two clumps will both grow too large in turn and split, producing four bad prion clumps. After another round of growth and breakage, those four will become eight, and so on. In other words, prions are a slow chain reaction. The end result is an exponentially growing number of prion vampires—and plenty of dead neurons and spongy holes.*

This prion theory also helped explain where kuru came from. Unlike kuru, Creutzfeldt-Jakob appears in ethnic groups worldwide. It usually starts when a gene mutates in some unlucky person's brain, and he begins to produce bad prion proteins spontaneously. Around 1900 some eastern highlander almost certainly came down with a form of Creutzfeldt-Jakob that attacked the cerebellum, and his equally unlucky loved ones consumed the brain. Prions are indeed immune to cooking and digestion, unfortunately, and can cross the BBB. As a result, the loved ones' brains got infected, and they died. The loved ones were in turn consumed, infecting still more people—who themselves died and were themselves consumed, and so on. Eventually, they started calling the killer kuru. Notice that it wasn't cannibalism per se that caused the outbreak; eating brains isn't inherently deadly. It was the bad luck of eating patient zero. Sadly, then, the very proteins that Fore women had so craved at the funeral feasts ended up killing them.

Since the 1980s prion research has swelled in importance. The outbreak of mad cow disease in the 1990s was basically a case of bovine kuru. British farmers were feeding the ground-up brains of cattle that had a prion disease to other cattle, who in turn infected the humans who ate them. (Not coincidentally, Prusiner won a Nobel Prize for prion research just after the mad cow scare, in 1997.) Disturbingly, some people might still have deadly bovine prions lying dormant inside them.

More recently, prion research has crossed over into mainstream neuroscience. The snarly protein plaques in kuru brains seem to grow and spread in much the same way as the snarly protein plaques that ravage the brains of people with Alzheimer's, Parkinson's, and other neurodegenerative diseases—first turning innocent proteins rogue, then clumping together into plaques that poison neurons and interfere with synapses. (There's even evidence that Alzheimer's plaques in particular require the presence of the normal prion proteins to do damage.) Thankfully, you can't "catch" diseases like Alzheimer's and Parkinson's. But if other scientists can build on this prion work and slow down or even cure these ailments—which affect more than six million people in the United States alone and will grow increasingly common as our population ages—more Nobels will surely follow.

Gajdusek himself had first suggested the kuru–Alzheimer link decades ago but didn't really pursue it. After winning the Nobel, in fact, he got more and more lethargic. There were still plenty of lectures worldwide as well as occasional jaunts to study diseases in places like Siberia. But having gained back the weight (and then some) he'd shed in New Guinea long before, he slowed down a lot, and spent more and more time at home with his adopted children.

Or rather, his adopted sons, since the vast majority of youths he'd surrounded himself with were male. A few colleagues, having noticed the pattern here—*All strapping lads, eh?*—began snickering about this and winking at each other whenever Gajdusek prattled on about "my boys." The FBI found the situation less funny.

As early as 1989, Maryland police had started investigating Gajdusek on charges of sexual molestation. The FBI got involved in 1995, when agents began scouring his published journals and field notes. A number of passages made them cringe. Passages describing various boys' pubic hair; passages about boys who, "with the slightest encouragement to their fondling, [go] searching in my pockets"; passages about him waking up on Christmas morning having "slept well again, like a bitch with her half-dozen pups lying and crawling over her"; passages about fathers who "smile and . . . indicate that I should let the boys play sexually with me." All of this was vague or ambiguous, though, and it all took place in New Guinea anyway. So the FBI began questioning his adopted sons, and finally found one who claimed that Gajdusek had had relations with him as a teenager in Maryland. (Other victims came forward later.) The young man agreed to call the seventy-two-year-old scientist, and during their conversation he asked him, "Do you know what a pedophile is?" Gajdusek allegedly answered, "I am one," then admitted to having sex with other boys. Gajdusek begged him to keep quiet, but the phone call was being recorded. Just before Easter in 1996, as a pudgy, jet-lagged Gajdusek pulled into his driveway one morning, returning from a conference in Slovakia on mad cow disease, a half dozen police cars whipped out of hiding, their red and blue lights screaming. Arrested and jailed on charges of "perverted practices," Gajdusek ranted from his cell, vowing to "pray to my pantheon of gods" for deliverance and attacking his accusers as "jealous, vindictive . . . probably psychotic." Eventually, though, he pleaded guilty and served eight months.

In subsequent interviews* Gajdusek more or less admitted everything: "all boys want a lover," he claimed, later adding, "and if I find them playing with my cock, I say good on you, and I play with theirs." He further defended himself by saying that the boys always approached him for sex, not vice versa, and that they came from a culture where sex between men and boys was appropriate, so no harm done. (An intellectual, he also invoked the widespread pederasty of classical

Greece.) In truth, the highlanders weren't the sexual libertines he claimed: they knew full well about other pedophiles who'd settled in their land and taken advantage of their culture, and they despised these men as perverts. Gajdusek also seemed willfully blind to the power he'd held over his boys in America as their guardian and master. Regardless, he never apologized, and fled to Europe after his jail term. He summered in Paris and Amsterdam, and wintered in northern Norway, enjoying the never-ending nights of winter solitude. He died, defiant and alone, in a hotel room in Tromsø, Norway, in 2008.

It's a complex legacy. Gajdusek was one of the outstanding neuroscientists of his era: he alerted the world to a brand-new brain disease, and his experiments on the brains of apes (along with Prusiner's crucial research) opened up a whole new realm of not-quite-living "biology." He also proved that infectious agents can crouch inside the brain for years before springing—a baffling notion then, yet one that foreshadowed the long latency of HIV. Moreover Gajdusek fought harder than anyone to help the victims of this cruel disorder, and it remains the only human disease besides smallpox ever eradicated: since 1977, 2,500 people have died of kuru, but none since 2005. The Book has probably had its last entry. Yet even as he fought to save Fore society, Gajdusek was apparently preying on its most vulnerable members. What's more, for all his sweat and blood, his work on the brain saved exactly no one; missionaries and patrols had largely stopped the cannibalism before he arrived, and every last person who caught kuru died. In the end, neuroscience proved impotent—and even today most Fore people remain convinced that sorcerers caused kuru.

But perhaps that's too bleak a view: the victims of kuru didn't die in vain. Basic biological research serves as the foundation for more and better work, and because of those victims' sacrifices we now know that kuru ravages the brain in ways tantalizingly similar to Alzheimer's, Parkinson's, and other plagues of old age. So perhaps the "world's rarest disease" holds the insight to preventing brain decay in human beings everywhere. If that proves to be so, the Fore will have gotten inside our brains as surely as they got inside the brains of so many scientists. And as neuroscience continues to expand its scope and map out how the tiny circuits in our brains give rise to higher-level drives and emotions, perhaps even the self-delusions and contradictory desires of someone like D. Carleton Gajdusek will start to make a little more sense.

CHAPTER SEVEN

Sex and Punishment

In addition to nerves and neurons, the brain also sends out signals via hormones. Hormones play an especially important role in regulating emotions, which provide a crucial bridge between brain and body.

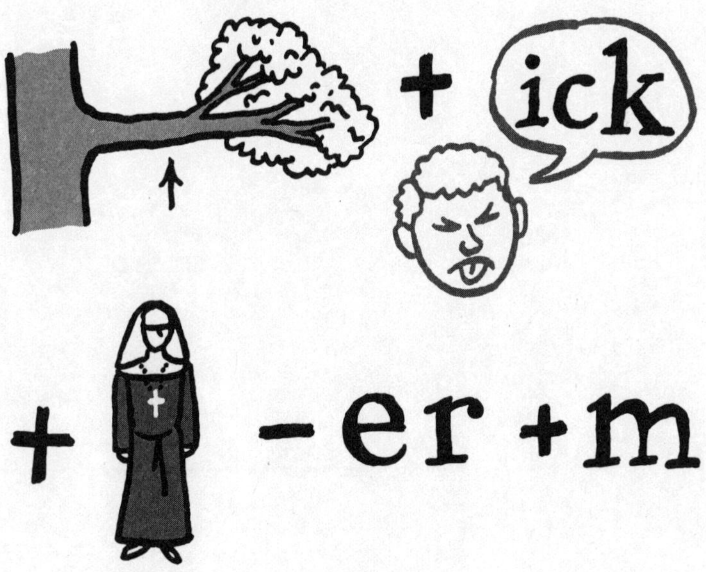

Neurosurgeon Harvey Cushing made quite a first impression. At 8 a.m. on New Year's Day 1911, a surgical resident named William Sharpe reported to Cushing's hospital in Baltimore for his first day of duty. If Sharpe expected an easy day—a tour, some chitchat, reviewing a few charts—Cushing had other ideas. Sharpe walked in to find Cushing wrist-deep inside a dog's skull, digging out the poor mutt's pituitary gland. Without preamble or introduction, Cushing handed Sharpe fifty dollars to bribe a priest and ordered him to rush to a Washington, D.C., funeral parlor. There, Sharpe was to remove all the endocrine glands—plus the brain, heart, lungs, pancreas, and testicles—of one of Cushing's patients, a giant. And since the funeral would start at 2 p.m., Sharpe had better scoot.

Neurosurgeon Harvey Cushing.
(National Library of Medicine)

At the funeral home the priest on duty collected his take and led Sharpe to the back room where the giant, John Turner, lay in a custom-built coffin. An illiterate brick-wagon driver, Turner had endured an excruciating growth spurt starting at age fifteen; he'd ended life, at thirty-eight, a seven-foot-three-inch Goliath barely able to walk. Sharpe asked the priest to help him lift Turner out of the casket, but the priest threw up his hands: it turned out that, contra Cushing, no one had permission to perform an autopsy. In fact, Turner's family opposed it. With no hope of moving the 343-pound giant, Sharpe had to flay him right there in the coffin. He undid the spinnaker-sized tuxedo jacket, and made the first incision around 11 a.m. Soon lost in his work, he barely heard the mourners gathering in the front parlor.

Around 1:00 p.m., Sharpe realized he'd made a tactical mistake: he should have started with the head. Of all the glands he had to secure, the pituitary—a small hormone factory, and Cushing's darling gland—was the most important. But it resided deep within the skull, so excavating it would require sawing through the skullcase, a noisy process. And by now, Sharpe could hear Turner's family getting restless in the other room—the priest had no good answer for why they couldn't see the body just yet. Sharpe sawed quickly, but Turner's overgrown skull turned out to be an inch thick in places. Sharpe soon heard fists pounding on the door, demanding to know what that noise was.

After popping the skull open, Sharpe peeled back the cobwebs of connective tissue around the brain and exposed the pituitary. The pituitary gland dangles beneath the brain like a dollop of tissue about to drip off. It's normally pea-sized but often swells grotesquely inside giants, due to tumors. Sharpe didn't have much time to examine Turner's gland, though, since the occupants of the front room had outright revolted by then.

Sharpe scrambled to stitch Turner up and gather his organs, and just as he finished, the dam burst. Luckily the priest (not wanting two dead bodies on his hands) had already called a cab for Sharpe. And as Turner's family spilled into the room, Sharpe dashed out the back door, hopped into his taxi, and yelled go. As the car sped away, a rock hurled in fury smashed against the trunk.

Back in Baltimore, Sharpe deposited the glistening viscera in a refrigeration unit. He called Cushing and retired early that night in a resident dorm, satisfied and proud. But just before dawn a hand shook him awake. Cushing stood over him, volcanic. "You missed the left parathyroid body!" he screamed. Sharpe tried to explain about the priest and the tuxedo and the rock—not to mention the fact that he'd never heard of the parathyroid body. Cushing cut him short and fired him.

Although inconsolable, Sharpe let two fellow residents treat him to breakfast in the hospital cafeteria. They explained that Cushing erupted like this all the time—he had poor emotional control. Halfway through their meal, the PA system crackled: *Paging William Sharpe. William Sharpe to Dr. Cushing's office.* Sharpe must have trembled as he approached the door—had he missed something else? But he found Cushing calm and levelheaded. He showed Sharpe a parathyroid gland and explained its function. Turning back to his work, he wished Sharpe a good day and never mentioned the incident again. (Sharpe, of course, remembered it for the rest of his life.)

Those twenty-four hours were classic Cushing. There was a legitimate medical question at stake; he sought out the answer ruthlessly; and somewhere along the way he lost his head. Even during his childhood in Cleveland, his family had called him Pepper Pot for his tantrums, and as an adult he blew up at nurses and residents almost daily. But Cushing's rage was easy-come, easy-go, and even those who got scorched couldn't deny his brilliance.

After attending Yale University—where he'd fallen out with his father because he insisted on playing baseball for the Bulldogs instead of focusing solely on academics—Cushing sailed through Harvard Medical School and secured a job at The Johns Hopkins Hospital in

Baltimore. He promptly derided Baltimore as dull, with row houses "as alike as streptococci" bacteria. But he found it convenient that his superior at Hopkins was a morphine addict,* since this left him unsupervised and free to experiment with new technologies. He used electricity to stimulate the brains of epileptic patients and built a device that displayed a patient's blood pressure and pulse automatically, so surgeons could tell at a glance if there was trouble. Cushing also started using X-rays in 1896, just a year after their discovery, to hunt for brain tumors and for bullets lodged in people's bodies. (Fortunately for him, he had too many other obligations to experiment much with X-rays; a less busy colleague died of radiation poisoning.)

Cushing eventually devoted himself to brain surgery, especially brain tumors. And although his exact and demanding temperament didn't endear him to his staff, those qualities did make him an excellent surgeon. Unusually for the time, he bothered distinguishing between different species of brain tumors and adjusted his approach accordingly. His ritualistic devotion to cleanliness reduced mortality rates for brain surgery from around 90 percent to 10 percent. And when a patient did die, he used the autopsy to check what he'd done wrong, which was almost unheard of. Finally, he had superhuman concentration. Brain surgeries can take ten hours or more—surgeons sometimes joke about the tumor growing back before they can sew the patient up. But Cushing could work on his feet indefinitely without flagging, and if he caught an assistant daydreaming, he'd snap—a flashback to Yale—"Eyes on the ball!"

By age thirty-two he'd built a comfortable practice, and could have cruised for the next few decades. But a devastating mistake changed the course of his career. In December 1901 he met a fourteen-year-old girl who was chubby, nearly blind, and sexually immature (no breast buds, no menarche). Cushing diagnosed excessive pressure inside her skull and opened her up to drain the excess fluid. She didn't improve, and soon died. The autopsy, to Cushing's chagrin, revealed a cyst pressing on her pituitary gland, a possibility he'd overlooked.

To be fair, most surgeons of that era would have overlooked it. Although not technically part of the brain, the pituitary sits on the brain's south pole, the neurological equivalent of Antarctica, so reaching it required a long, invasive surgery. Even when surgeons did reach it, they hesitated to touch it, since it nestles right near the optic nerves (hence the girl's near blindness). And no one knew what good touching it would do anyway, since its function remained a mystery. Some called it an evolutionary relic, like an appendix, while others linked it to a baffling array of diseases: face and hand deformities, obesity, skin problems, even (contradictorily) both gigantism and dwarfism. Given its obscurity and inaccessibility, most surgeons preferred to pretend it didn't exist. Cushing was not most surgeons, and determined to get to the bottom of it.

However puzzled, the more Cushing read about the gland, the more it intrigued him — especially its association with growth disorders. As a lad, whenever the circus had swung through Cleveland, Cushing had raced to see the sideshows, gaping for hours at the giants, dwarfs, fat ladies, and other "prodigies." Pituitary work rekindled this illicit fascination, and in the early 1900s — under the guise of research — he started visiting giants and dwarfs up and down the East Coast, as well as freaks in traveling circuses, and taking detailed medical histories.

A few, like John Turner, bristled at this intrusion. But most remembered their slight, dapper doctor fondly and even let him visit their homes. Cushing's most memorable consultation took place late one night in Boston, when he rendezvoused with the World's Ugliest Lady in her private boxcar outside a train station. Because the train yard was a labyrinth, Cushing requested a guide, and his Virgil turned out to be a three-foot dwarf brandishing a lantern. The flickering oil light made for some eerie visual effects on their journey, and Cushing later remembered that once or twice as they wandered, the dwarf seemed to transform into "a hobgoblin" before his eyes. At the Ugly Lady's boxcar, a giant lifted the dwarf onto the platform, then helped lift the five-foot-seven-inch Cushing. Inside he found the circus's menagerie of freaks splayed about on couches, including a "half-lady" with no legs. Cushing remembered thinking he'd stumbled into a fairy tale, and throughout his interview he felt a smile twitching his lips, even as their tales of suffering stung his eyes with tears.

After studying dozens of freakish medical histories, Cushing determined that the pituitary gland's primary job was communication. We normally think of the brain communicating with the body through nerve impulses; that's how the brain creates movement, for instance. But the brain has other ways of issuing commands, through chemicals such as hormones. It's just that instead of secreting hormones directly, the brain sometimes outsources that job to glands. And there's no more important gland for hormone regulation, Cushing declared, than the pituitary "master gland." It's a veritable hormone factory, with half a dozen different types of cells, each of which secrete distinct hormones. And the over- or underproduction of any one hormone can cause a distinct disease, which explains why the pituitary was linked to so many different ailments.

Thanks to his obsession with giants and dwarfs, Cushing zeroed in on one hormone in particular, growth hormone. If a child gets a pituitary tumor, the cells that secrete growth hormone often start to proliferate. The gland then pumps out too much growth hormone, and the child shoots up into a giant. In contrast, if a nearby cyst crushes these cells, the gland produces too little growth hormone and renders the child a dwarf. It turns out, then, that there was no "contradiction" with the pituitary: problems with it could indeed produce both dwarfs and giants.

Cushing also determined that pituitary disturbances after childhood caused different ailments. People who reach an adult height obviously won't shrink back into dwarfs if their hormone-producing cells die. But they can regress sexually, becoming apathetic about sex and packing baby fat onto their faces and bellies. Men's genitals might retract, too; females might stop menstruating. Similarly, a revved-up pituitary gland won't make adults taller, since the growth plates in their arms and legs have fused. But a flood of growth hormone in particular can cause acromegaly, a condition where people's hands, feet, and facial bones thicken* and their eyes bulge out as if they're being strangled. Cushing's "ugly lady" in the boxcar had this unsightly disorder.

Excited by his discoveries and eager to expand his neurosurgical repertoire, Cushing began operating on pituitary cases in 1909. His first patient was John Hemens, a farmer from South Dakota whose hands and feet had grown several sizes in adulthood. (Many people with pituitary trouble have to buy bigger gloves and boots every few years.) Hemens's face had swelled grotesquely, too: his tongue and lips grew so thick he could barely speak, and gaps had opened up between his teeth where his jawbone had expanded. Head to toe, it looked like classic acromegaly, and Cushing decided to remove part of the pituitary. He knocked Hemens out with ether, then entered his skull through an omega-shaped incision (Ω) just above the nose; after wriggling his instruments into the brainpan, Cushing lopped off a good third of the gland. Hemens woke up minus his sense of smell, but his

Left: John Hemens, a farmer from South Dakota who suffered from acromegaly, before developing symptoms. Right: Hemens just before Harvey Cushing operated on him.

long-standing headache and eye pains had disappeared, and the swelling in his face and hands soon subsided. Cushing declared him cured.

The relief didn't last, unfortunately; many of Hemens's symptoms returned within a year. But the result nevertheless encouraged Cushing: no one had ever alleviated a pituitary disorder before, even slightly. So Cushing kept pushing forward, doing bolder operations. In 1912 he even dared transplant a pituitary gland from a dead infant into a comatose Cincinnati man whose pituitary had been destroyed by a cyst, in a last-ditch effort to save the man's life. The patient died without regaining consciousness, and the case caused a scandal when a newspaper reported, in a howler, that Cushing had actually transplanted the baby's entire brain. Even this setback couldn't deter Cushing, and he kept developing new treatments.

Soon enough, a colleague putting together a new book asked Cushing to submit an eighty-page chapter on the master gland. Cushing submitted eight hundred pages. He later condensed that into his own book, *The Pituitary Body and Its Disorders.* Frankly, the book had some shortcomings. Always obsessive, Cushing had started attributing any and all disorders of unknown origin to this "potent mischief-maker," and had therefore included some dubious cases. Nevertheless, the book deserved its fame. Before it, doctors mostly ignored giants, dwarfs, and fat ladies as inexplicable "prodigies." And if doctors did dare to treat a hormone imbalance, they usually (especially with women) simply tore out the sex organs and hoped for the best. Cushing provided a more rational, more humane alternative, and provided the first real relief that most victims had known.

Beyond its medical merits, the book became famous for another reason—its haunting photographs. Thanks to his love of technology, Cushing had long ago started "kodaking" his patients to document their deterioration. Indeed, he was an early proponent of "before" and "after" comparison shots—except that he inverted the modern gimmick, since his patients always looked worse in the second picture. Dandies in three-piece suits suddenly had shrunken genitals and so

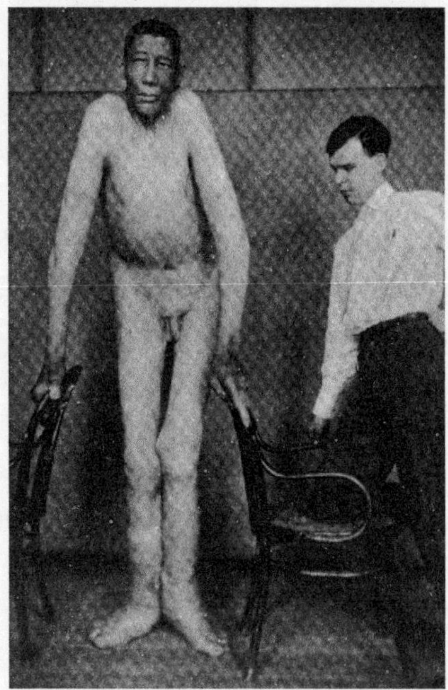

Pituitary giant John Turner. Even decades later, Cushing's assistant, right, still got stopped by people at medical conferences who recognized him from this photo.

much belly fat they looked pregnant; elegant mademoiselles suddenly sported humpbacks and mustaches. Perhaps the saddest photo showed the giant John Turner propping himself up on two chairs, his face alarmingly ruddy and his legs looking rickety enough to collapse beneath him. (A five-foot-eight-inch assistant of Cushing's, who happened to be standing next to Turner in this picture, often got stopped at medical conferences even decades later because people still recognized him from that photo.)

In some ways the book combined the artistic innovations of Vesalius with Velázquez's interest in human deformity. But whereas Velázquez lent his subjects dignity, Cushing's photographs make for uncomfortable viewing today. There's something almost ruthless about them. Turner and most other patients were stripped naked; none had black bars over their faces or genitals, and most didn't even try to smile. They call to mind that old superstition about cameras stealing people's souls. Certainly, Cushing cared for his patients as people. He kept up correspondence with hundreds of them, and excoriated *Time* magazine once when it mocked some circus freaks (including Cushing's "ugly lady") in a column. At the same time, Cushing wasn't above making a crack himself in private, and he certainly capitalized on the public fascination with freaks to cement his own fame.

Already by 1913 Cushing had parlayed his book into an appointment at Harvard, and despite his punishing schedule he took on even more work throughout the 1910s and 1920s. He wrote a biography of his mentor, which won a Pulitzer Prize in 1926. Meanwhile he began a massive project—whole tables in his house were covered in paraphernalia—to hunt down every surviving manuscript and original likeness of Vesalius, whom Cushing considered his spiritual forefather. (Cushing even tracked down some rare first editions of *Fabrica*—one, oddly, found in a Roman blacksmith's shop.) He got so wrapped up in this and other work that he didn't notice the stock market crash of 1929 until months later, when patients stopped trickling in. He nevertheless celebrated removing his two thousandth brain tumor* in 1931. Along the way—always confident in his own abilities—he also operated on his family members a few times, removing two of his children's appendixes and a tubercular growth from a daughter's neck.

Decades of nonstop work finally wore Cushing down, and his health deteriorated further after he himself was diagnosed with a small brain tumor in the late 1930s; he eventually died of complications from a heart attack in 1939. Again, not all of Cushing's early work has held up. But more than anyone else of his era, he illuminated how the brain's glands work (or fail to), and he laid crucial groundwork for understanding how the brain influences the body. Even as Cushing lay dying, in fact, other scientists were expanding on his insights and linking his master gland to another major brain-body system, the system that produces human emotions.

~

Appropriately enough, the modern study of emotions began with one man's rage. In 1937 Cornell neuroscientist James Wenceslaus Papez (rhymes with "drapes") got wind of a new research grant being offered; the money would help scientists study how emotions work inside the brain. Papez thought the underlying implication of the grant—that scientists knew nothing about the neuroscience of emotions—insulted some colleagues of his who'd already illuminated key aspects of the field. So in what he later described as "a fit of spleen," Papez wrote a paper outlining the current state of knowledge. But the whole of his paper far exceeded the sum of its parts.

Papez drew on cases of brain damage where people's emotions went into under- and overdrive. For instance, lesions in the thalamus, a cluster of gray matter deep inside the brain, can cause spontaneous laughing or crying. In contrast, when another internal structure, the cingulate gyrus, is destroyed, people become emotionally vacant. Perhaps most powerfully, Papez cited cases of rabies. Because swallowing can cause painful neck spasms, many rabies victims actually fear water—it terrifies them. (The inability to swallow also contributes to foaming at the mouth because of excessive saliva.) Rabies produces aggression as well, by attacking certain gray-matter clusters in the brain. The resulting aggression makes dogs, raccoons, and other animals more volatile and therefore more likely to bite and pass on the virus. Human victims lash out in similar rages, and hospital staff often have to lash patients to a bed.

In all these cases, brain damage led to either exaggerated or blunted emotions, and Papez soon realized—his big insight—that these damaged structures must work together in some sort of "emotion circuit." Scientists later named this circuit the limbic system. The word "limbic" comes from the same Latin word that gave us limbo, and the limbic system does indeed serve as a transition between the brain's upper and lower regions. As one neuroscientist put it, "like the limbo of Christian mythology, the limbic system is the link between the cortical sky and the reptilian hell."

Scientists have bickered about what structures do and don't belong in the limbic system practically since Papez finished the first draft of his paper. This confusion arises in part because different people mean different things by "emotion"—a subjective feeling, a flood of hormones, a physical response or action. Things also get confusing because different emotions activate different brain structures. Finally, the limbic system interacts with so many other brain regions that it's hard to draw a boundary around it.

Middle Brain/Limbic System

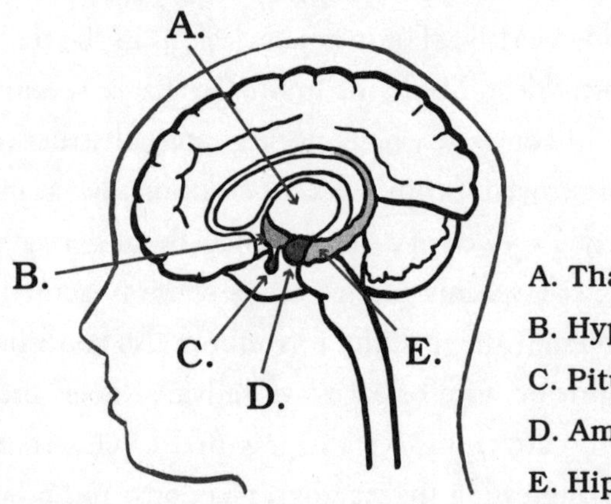

A. Thalamus
B. Hypothalamus
C. Pituitary gland
D. Amygdala
E. Hippocampus

Still, the limbic system does have a few core components, many housed in and around the temporal lobes. In general they work like this. Say you see something scary, like a tiger. Sights and sounds get filtered through the thalamus, a double-lobed structure smack in the center of the brain. The thalamus takes a first pass at this sensory input—*claws, teeth, growling, hmmm*—and splits the data into multiple streams for further processing. One stream flows into the hippocampus, which helps form and access memories. Another stream splits into two branches, one of which flows straight into the amygdala, the other of which detours into the frontal lobes before also arriving at the amygdala. (More on the amygdala in a moment.) By now the brain has a good grasp on the tiger, so it's time to alert the body. To do so, the amygdala pings the hypothalamus,* one of the busiest beavers in the body: it takes full or partial responsibility for all sorts of bold-faced Biology 101 terms like metabolism, homeostasis, appetite, and libido, among other things. (Biologists summarize these hypothalamic duties as the "four F's" of animal behavior—feeding, fleeing, fighting, and, well, sexual congress.) Finally, neurons in the hypothalamus rouse Cushing's pituitary gland, which in turn releases hormones into the bloodstream that cause us to sprint, shiver, wet ourselves, or otherwise feel emotions viscerally.

The limbic system reaches far and wide, and interacts with many other parts of the brain and body. It pings our facial muscles to produce blushes, growls, grins, and grimaces. It can take feedback from the face as well. The mere act of smiling, for instance, can send uplifting hormones surging through us, brightening our moods. (The brain associates smiling with a good time, so once the smile circuits flash on, the good-mood circuits often flash on, too. Something similar happens with frowns and feeling crummy. Wire together, fire together. And on the flip side, hindering people's facial expressions—with Botox, for instance, which paralyzes facial muscles—can actually dampen emotions like rage.) As for mental phenomena, limbic structures work with the frontal lobes to produce rich emotional experiences such as euphoria and melancholy and lust, the swells and highs that make us feel fully alive. Part of the horror of lobotomies was that they severed the links between the frontal lobe and limbic system, blunting or even destroying emotional experiences.

Ironically enough, Papez's paper on the limbic system—written to make the grant providers look foolish—provided exactly what the grant was seeking, an outline of how emotions work. But for all his brilliance, Papez missed one crucial aspect of the limbic system: his 1937 paper skipped right over the amygdala. Named for their shape (from the Greek for "almond"), the two amygdalae sit deep inside the temporal lobes. They jump-start the startle reflex; process smell, the one sense that skips the thalamus; and help determine which things around us are worth paying attention to. Indeed, some neuroscientists, riffing on the *what* and *where* streams in vision research, have declared the amygdala and other nearby structures the *so what* stream. *I see this,* your brain says, *but should I bother caring?* The amygdala helps make that call.

If we should care, the amygdala also takes the next step and helps mount an appropriate response, especially if that response involves fear. In fact, the amygdala is often called *the* fear spot in the brain. That's simplistic—the amygdala processes lots of emotions, including happy ones—but there is some truth to it. Having a structure to scan for frightening things is mostly good, since it steers us clear of fanged animals, dark spaces, clowns, and so on. But like any other brain part, the amygdala can malfunction, making people feel fearful all the time. They see threats where none exist, and might go postal if pushed too far.

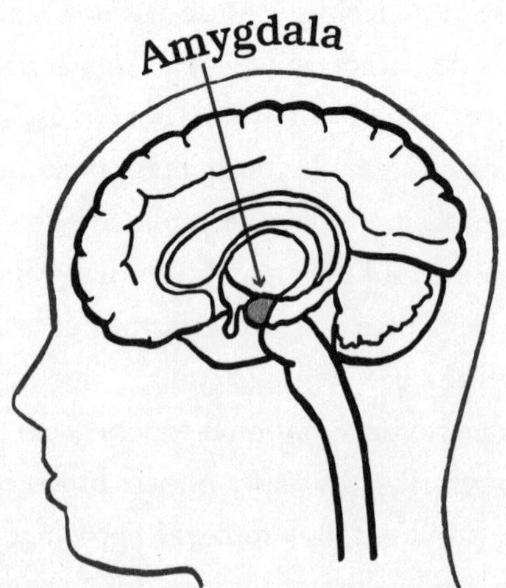

Conversely, as research on a woman named S.M. shows, amygdala damage can also lead to the opposite problem—an alarming lack of fear. As a child S.M. responded normally to scary things. One midnight she followed her brother into a graveyard, and when he leapt out from behind a tree, she screamed. She also got cornered by a Doberman once and felt her heart quake and her gut seize up—typical fear responses. But around age ten she began suffering from Urbach-Wiethe disease, a rare disorder that petrifies and kills amygdala cells. Within a few years she had two "black holes" where her amygdalae should have been. She hasn't felt a lick of fear since.

Studies involving S.M. are actually a hoot to read, since they basically consist of scientists dreaming up ever-more-elaborate ways to scare her. For instance, doctors once drove her to an exotic pet store that carried snakes. On the ride over she claimed to hate snakes, but as soon as she arrived she practically yanked the serpents out of the store clerk's hands to play with them. She even tried stroking the snakes' tongues (snakes do *not* like this), and asked fifteen separate times to pet some of the venomous snakes on hand. Her doctors also ran her through a haunted house—an old lunatic asylum with plenty of creaking doors and shadowy corners for monsters to leap out of. Five strangers, all women, toured with S.M. and served as effective controls. They screamed every few seconds, but S.M. was always dashing ahead to find out what was next. At one point she jabbed a monster—an actor on duty—in the head because she wanted to know what his mask felt like. *She* ended up scaring *him.*

To test whether S.M. was simply blunted (i.e., dead to all emotion), her doctors had her examine photographs of people making various faces. She could read most emotions just fine, but fear didn't register. Similarly, when viewing a range of movie clips, she reported feeling sad, surprised, mirthful, and disgusted at appropriate times but barely blinked during *The Shining* and *The Silence of the Lambs.* What's more, family members report that, if anything, S.M. gets overly emotional sometimes, excessively sad and lonely. This pattern makes sense, because while other emotions can sidestep the amygdala, fear cannot: to feel fear, the amygdala needs to rouse us.

Lest you think these experiments sound contrived—S.M. didn't face any real danger inside the haunted house, after all—consider the "knife incident." While walking home alone one night, she cut behind a church and into a public park. A man she described as "drugged-out" yelled at her, and yet she walked over to him without hesitating. He grabbed her, whipped out a knife, and held it to her throat, hissing, "I'm going to cut you, bitch!" She didn't struggle. Instead, she listened to the church choir wrapping up rehearsal, then murmured something about God's angels protecting her. The man, weirded out, let go. At this point, rather than spring and sprint for her life, S.M. simply walked away. She even returned to the park the next day. S.M. has also been held at gunpoint and once almost died during an episode of domestic abuse. But despite feeling upset, the words she used to describe these incidents made them sound like inconveniences or annoyances. Fear never came up.

Critics of S.M.'s case have argued that her behavior sounds less like a lack of fear and more like a lack of common sense—that the black holes in her amygdalae were really just holes in her head. But knowledge of the limbic system negates that criticism. Notice that when S.M. saw a snake or some other danger, she didn't just shrug—she was dying to get her hands on it. This makes sense biologically. If you saw a viper in the wild, you wouldn't want to get distracted; best to pay close attention. On some level, then, S.M.'s brain did recognize frightening things, because she fixated on them. Her brain simply couldn't mount the subsequent emotional response to get her the heck outta there. So saying that S.M. lacked common sense misses the point. When it comes to detecting danger around us, fear *is* common sense. Fear bestows that common sense in the first place, and you can't have one without the other.

~

Although processed within the limbic system, emotions often spill over into other areas of the brain, in ways subtle and surprising. Some blind people with visual cortex damage—who have no conscious visual awareness of anything around them—can still read emotions on other people's faces. That's because the optic nerves, in addition to routing data to the conscious mind, also route data to the limbic system along secondary, subliminal tracts. So if the conscious pathway gets damaged but the unconscious, limbic one remains intact, blind people will still respond to smiles, scowls, or quivering lips, all without realizing why. They can even catch yawns* from other people.

Similarly, the limbic system can bypass certain kinds of paralysis. People who suffer a stroke in the brain's voluntary movement centers often have trouble smiling on command: the right side of their mouths might perk up, while the left side droops pitifully. Tell them a joke, though—something that engages a genuine emotion—and they'll often brighten right up with full, radiant, symmetric smiles. That's because the limbic system connects to the face through different axon channels than our voluntary motor centers do; the limbic brain can therefore still move the facial muscles whenever we ourselves feel moved.

(What's more, the limbic system and voluntary motor centers actually move different sets of facial muscles—and therefore produce different-looking smiles. This divergence explains the difference between genuine smiles and fakey, say-cheese smiles in photographs. People have trouble faking other genuine expressions, too, like fear, surprise, or an interest in someone's pet stories. To overcome this limitation actors either drill with a mirror and practice conjuring up facial expressions à la Laurence Olivier, or, à la Constantin Stanislavsky, they inhabit the role and replicate the character's internal feelings so closely that the right expressions emerge naturally.)

The limbic system, and the temporal lobes generally, are also closely tied up with sex. Scientists discovered this connection in a roundabout way. In the mid-1930s a rogue biologist named Heinrich

Klüver started some experiments with mescaline (a.k.a. peyote), a hallucinogenic plant. He kicked these experiments off during a summer holiday in New Hampshire when—bored out of his skull and lacking any lab animals—he solved both problems in one go by giving mescaline to a farmer's cow. Whether Klüver injected the beast with a syringe or let it nibble dried peyote buttons from his hand isn't known. What is known is that the cow croaked, and the farmer got hopping mad. Despite this inauspicious start, Klüver decided to sample mescaline himself, and nearly died. Undeterred, he began still more experiments, on monkeys, when he returned to his University of Chicago lab.

Around 1936 Klüver developed a theory that all hallucinations originate inside the temporal lobes. To test this idea he had a colleague, neurosurgeon Paul Bucy, remove the temporal lobes of a few monkeys. (This excavation also tore out key limbic structures.) The experiments failed—the monkeys still got high—but the duo did notice some odd side effects. For one, the monkeys lost the ability to recognize objects, even food. The monkeys also developed an oral fixation. The scientists determined this by scattering peppermints, sunflower seeds, and banana slices on the floor; they also scattered nails, lint, combs, eggshells, tin foil, cigarette ash, and seemingly whatever else they could scrape from their desk drawers. Instead of going straight for the treats, the monkeys methodically picked up every last item and licked or bit it, a trait now called hyperorality. The monkeys even subjected baby rats and feces to the taste test. Equally disturbing, the monkeys turned into sex fiends. They masturbated themselves raw and rubbed their genitals on any animate being in sight. One poor fellow, who combined the worst of hyperorality and nymphomania, had to be put down because, unable to recognize it, he kept biting his own penis.

Nowadays a neuroscientist would blame the monkeys' inability to recognize food on the destruction of the *what* stream within their temporal lobes. But their lack of a functioning limbic system also contributed to their bizarre behavior, since one purpose of emotions is to help animals appreciate and react appropriately to objects. In short, in a brain with a functioning limbic system, the amygdala's *so what* circuit will "tag" different objects with good or bad emotions. When we encounter those same objects later, the tag tells us whether to run, smile, fight, or approach. Monkeys respond appropriately to bananas, for instance, because bananas assuaged their hunger once and gave them a shot of sugar. This in turn flooded their brains with dopamine, a neurotransmitter linked to rewards. So upon seeing bananas again, the monkeys repeat the steps—approaching and eating—that led to that good feeling before. Conversely, they shy away from fire and snakes because those things are tagged as frightening, and they shun feces because they're tagged as disgusting.

Now, imagine that those tags have vanished. No one thing would seem any more desirable or repulsive or terrifying than any other thing—which is exactly what happened to Klüver and Bucy's monkeys. Minus a limbic system, bananas and lint and poop nuggets all seemed like potential food, and no matter how many times they grabbed lit matches or humped a technician's leg, they never hesitated to do it again. And if you think that all this failure and futility frustrated the monkeys no end, you'd be wrong. Because they lacked a limbic system, they never got upset about repeating the same damn mistake over and over. In fact they never really betrayed any emotions, ever. No joy, no resentment, no anger, no nothing. Even when a rival monkey nearly bit through one de-lobed fellow's hand—something no self-respecting primate would tolerate without a brawl—the bitten one just yanked it away and sauntered off.

Klüver and Bucy studied monkeys, but human beings with limbic damage display many of the same traits, a disorder now called Klüver-Bucy syndrome. Like the bitten monkey, one symptom of Klüver-Bucy is "pain asymbolia," which leaves people indifferent to physical pain. They can recognize intellectually that getting a hand crushed or having a needle break off beneath their skin should hurt, but because the pain lacks any emotional impact they don't get worked up about it. Klüver-Bucy victims also become hyperoral, and doctors have caught them chomping down on soap, catheters, blankets, flowers, cards, pillows, glass thermometers, and everything else in their hospital rooms. One victim suffocated while trying to swallow an Ace bandage.

As for sex, human beings often respond differently to brain damage than monkeys do. Seizures that send volleys of electricity into the limbic system, for example, can actually depress sexual appetite, leading to impotence and Dead Sea–level libidos: some epileptics have never had an orgasm in their lives. (In contrast, rabies infections can lead to spontaneous ejaculations—up to thirty per day.) Temporal lobe lesions can flip people's sexual orientations from gay to straight (or vice versa), or redirect their sexual appetites toward inappropriate things: common side effects of Klüver-Bucy include zoophilia, coprophilia, pedophilia, and -philias so idiosyncratic they don't have names. In 1954 three scientists published a report about an epileptic named L.E.E., a thirty-eight-year-old carpenter who, ever since boyhood, had been sneaking into bathrooms not with nudie magazines but with safety pins. After withdrawing the safety pin from his pocket—the shinier the better—he stared at it with increasing titillation for a minute, then became glassy-eyed. He hummed, sucked his lips, and went rigid; his pupils dilated. It's unclear if L.E.E. actually climaxed this way, but he didn't care. He claimed that his shuddering, groaning, mini-seizures actually trumped orgasms, providing much more pleasure. (And titillation wasn't the only benefit of his fetish. When he demonstrated it for military recruiters during World War II, they couldn't reject him fast enough.) Still, the fetish strained his marriage: by his thirties he couldn't get an erection during foreplay, and his wife was threatening to walk out. Only when surgeons removed a three-inch shank from his temporal lobe did he find any relief; afterward, he and the missus enjoyed connubial bliss.

In truth, though, L.E.E. got lucky: all too often, removing temporal lobe tissue to stop one problem only introduces another. Most commonly, if the offending tissue suppressed someone's sex drive, removing it can cause his libido to skyrocket, and he'll get horny—beyond horny. One surgical patient started getting erections that lasted several hours, and within seconds of ejaculating he would roll over onto his wife for more sex; no amount of coitus could sate him. As you can imagine, this isn't easy for spouses to take, and some have even pulled their loved one's brain surgeons aside and requested a temporal lobectomy for themselves, so they can "keep up." Parents have it rough, too. Children as young as three (who have their temporal lobes removed to control intractable epilepsy) might start flashing their genitals and thrusting their little hips. One twenty-four-year-old lobectomy patient started begging sex from strangers, neighbors, and family members, and if they weren't ready to go, she'd start masturbating wherever. Hospitalized once for a seizure, she escaped from her room within half an hour. Her doctors found her beneath the sheets of an elderly man who'd just had a heart attack, her head bobbing up and down—combined hypersexuality and hyperorality. (As one commentator said, "One person's syndrome was another's lucky day.") Interestingly, she never remembered her "episodes" afterward.

Aside from our vision, motor, and sex centers, one final region that interacts with our limbic circuits is the frontal lobes, which help soothe and placate our more primal emotions. That's not to say that the frontal lobes quash emotions completely. A mysterious rustling in a dark forest will always sound alarms in the amygdala and send a pulse of fear through us. But instead of letting that fright overwhelm us, the frontal lobes, which are more discriminating and less reactive, help defuse and master it somewhat. This frontal-lobe influence also allows humans to have a more nuanced repertoire of emotions than other animals, who usually fall back on inflexible and stereotyped reactions.

All that said, we shouldn't congratulate ourselves just yet about being such smart, rational Vulcans. We've all been overwhelmed by fear or anger at times. And as the story of a man named Elliot shows, even the vaunted "higher reasoning" of our frontal lobes owes a heavy debt to raw emotion.

Elliot was good people. A caring husband who'd married his high school sweetheart; a father of two; head accountant at work; a solid member of his Iowa community. But in 1975, at age thirty-five, he started getting excruciating headaches, so piercing he couldn't think. Brain scans confirmed the worst: a baseball-sized tumor lodged above and behind his eyes. Actually, the tumor itself wouldn't have done much harm, except that in the enclosed space of the skull, it was crushing his frontal lobes. When surgeons opened him up, in fact, they had to remove whole swaths of damaged tissue from the prefrontal area, a region in the very forefront of the brain that contributes to planning, decision-making, and personality traits. A much different Elliot woke up from surgery.

This Elliot couldn't do something as simple as choose where to eat dinner. Before picking a restaurant he had to weigh the prices, menu, atmosphere, proximity to home, and quality of the waitstaff—then drive to every single option to see how busy it was. And after all that, he still couldn't decide. In fact, no matter what the decision was, Elliot went round and round, dithering and dallying and never making up his mind. Imagine every piddling choice in your life—what tie to wear *(stripes or Snoopy? Hmmm);* what side dish to choose *(soup or salad? Hmmm);* what radio station to listen to *(smooth jazz or classic country? Hmmm)*—being subject to such intense and futile scrutiny.

Elliot fared no better professionally. Although punctual before, this Elliot needed hounding every last morning to get his butt to work. He might spend hours shaving or washing his hair, because he simply didn't care about arriving on time, or at all. Not that he was much use at work. Even though his math skills remained intact, he couldn't manage his time and got distracted by meaningless tasks. He might fritter away a whole morning deciding how to file some documents, for instance. By color? Date? Department? Alphabetically? *Hmmm.* Hour after hour of shuffling and reshuffling ensued, with

Elliot entirely indifferent to both the hours he was wasting and his boss's glares. The inability to see the big picture is a common consequence of prefrontal damage, and often leaves a victim unable to get beyond step one or two and complete a task.

Finally, Elliot's personal life unraveled. After his inevitable firing, he drifted between jobs, working in a warehouse one season, preparing tax returns the next. Neither gig lasted. A shady local character then convinced him to invest his nest egg in a house-building scheme. When the investment got wiped out, Elliot shrugged. He fooled around on his wife of seventeen years, too, and after their divorce he married a prostitute, a union that fell apart six months later.

The strange thing was that Elliot's memory, language, and motor skills remained intact, and his IQ remained in the 120s. He could discuss economic news and domestic policies in detail, as well as foreign affairs in Poland and Latin America. Most oddly, he could reason just fine in controlled settings. When presented with hypothetical scenarios about people's social lives and asked to predict which choices would lead to happiness and which to ruin, Elliot could predict that, say, marrying a hooker maybe wasn't the best idea. Nevertheless, he never bothered steering clear of such disasters in his own life. Why? Because disasters didn't bother him—he didn't sweat the big stuff.

Neuroscientist Antonio Damasio has written about Elliot at length. And although it's a subtle case, Damasio argues that Elliot's lack of emotional distress provides the key to understanding him. The human brain normally has strong neural connections between the emotional limbic circuits and the rational prefrontal areas, and we normally think about this relationship in master-slave terms, with the rational brain defusing our emotions and suppressing our impulses. But there's more to their relationship, Damasio says. Emotions also advise the rational brain, and allow it to take past experiences into account when making decisions. Emotions do this by, once again, tagging any choices we face, A or B, as good or bad, based on how similar choices turned out in the past. Sometimes these tags even produce "gut feelings," allowing the wisdom embodied in our bodies to bend back and influence our minds. Overall, Damasio argues, this is the basic evolutionary purpose of emotions: to nudge us toward "good" options by associating them with positive feelings and deter us from "bad" options by stirring up unease.

Elliot's tumor destroyed key connections between his prefrontal lobes and limbic center, so the Socratic dialogue between reason and emotion never took place. This doomed him in small life choices, because choosing paisley over plaid or Chinese over country buffet depends little on ratiocination. Instead, it's emotion—*what do I feel like?*—that nudges us toward A and away from B. And minus emotions, Elliot's frontal lobes literally could not decide. Logic cannot make us make a choice.

The lack of dialogue also doomed him in big life choices. Elliot's prefrontal damage wouldn't have changed his basic drives and appetites, whether biological (e.g., for sex) or cultural (e.g., for money). In fact, those urges were probably normal, no stronger or more perverse than the urges the rest of us have. It's just that in most human beings, the prefrontal area curbs these urges and redirects them in socially appropriate ways; that's one of the most important jobs of the frontal lobes. And without the frontal lobe influence, Elliot's immediate drives and impulses *(me want sex)* always won out, hijacking his mind and forcing him into the same gimme-now decisions an animal would make. Perhaps worse, his lack of frontal–limbic interaction ensured that he couldn't tag his decisions emotionally as good or bad and thereby avoid similar mistakes in the future.

Notice that Damasio's work here* upends the traditional thinking about reason "versus" emotion. Emotions can cloud our reasoning, certainly. And again, abstract reasoning can proceed quite well without emotion: when presented with hypothetical scenarios in the lab, Elliot could foresee the disastrous consequences of certain decisions. It's the next step that flummoxed him. To most of us the next step seems so obvious that it feels stupid to spell it out—avoid the decisions that lead to ruin, you idiot. Even after outlining all the negative consequences, however, Elliot would typically grin and admit that in real life, "I still wouldn't know what to do." It seems daft, the antithesis of common sense. But as we learned with regard to fear, it's probably emotions that produce common sense. And when none of his possible options were marked as scary or dangerous or joyful, Elliot floundered. Overall, then, while reasoning without emotions might seem ideal in the abstract, in practice—in Elliot—it looks like irrationality incarnate. This is one of the hardest truths of neuroscience to swallow: that no matter how much we want to believe otherwise, our rational, logical brains aren't always in charge. We crowned ourselves *Homo sapiens,* the wise ape, but *Homo limbus* might have been more apt.

~

Elliot's struggles bring up another point, a deeply ethical point, about how much responsibility we have for our actions. Brain damage can release some dark and primitive impulses, and Elliot had particular trouble choosing right over wrong in the grip of an immediate temptation. Imagine, though, that instead of investing poorly or ruining his marriage, Elliot had embezzled money or murdered his wife. Our entire legal system is built on the premise that people who understand right and wrong are accountable for their actions. But in the light of neuroscience, jurists have struggled with in-between cases—cases where someone understands right and wrong, and even understands the need to choose right over wrong, yet lacks the power to.

One case involved a teacher from Virginia. Although he had a penchant for pornography, he lived a fairly vanilla life until around age forty, when he started soliciting happy endings at massage parlors. More troublingly, he started collecting smutty videos of underage girls, and even though he tried to resist these urges, he soon approached his eight-year-old stepdaughter for sex. She informed his wife, who found the kiddie porn stashed on his computer. Arrested and tried, the man couldn't explain himself in court. He'd never lusted after children before and knew he shouldn't now, but he couldn't help it. Citing the lack of even a whiff of criminal activity in the man's past, a judge sentenced him to attend rehab in lieu of jail. Things didn't work out there: he kept asking the nurses to mount him; even after pissing himself one afternoon, he kept propositioning them, the Romeo. The center expelled him, so off to prison. But the night before his sentence began, he complained of a massive headache and checked into a hospital. You guessed it: he had an egg-sized tumor in his brain.

Was this a coincidence? Statistically speaking, some percentage of pedophiles will have brain tumors, tumors unrelated to their vice. And if not a coincidence, did the tumor simply release his dark desires, or did it create desires that didn't exist before?

When surgeons removed the tumor in December 2000, the pedophilia disappeared. For a while. The man started pursuing children again the next October. But because his headaches returned as well, his doctors scheduled another brain scan, and sure enough, as so often happens, the surgeons had missed a tiny root of the tumor, and it had grown back like a pernicious weed. When surgeons removed it a second time, the pedophilia once again disappeared. This seems to argue that the tumor somehow caused the pedophilia. But again, we don't know whether the tumor simply released a pent-up desire or whether it actually changed his mental makeup. This is not an isolated case, either. One study from 2000 found at least thirty-four men whose pedophilia emerged after tumors, trauma, dementia, or other insults to their gray matter. To be sure, most pedophiles do not suffer brain damage, but clearly some percentage do.

There are a few key factors to weigh when judging whether or not you can blame criminal behavior (or other outré activities) on brain damage. One is the presence of additional trouble. When examined, the Virginia man failed multiple neurological tests, being unable to keep his balance or write a legible sentence; he also showed the same snout reflex that kuru victims did. Equally important considerations include how quickly the new behavior emerged and the contrast between the patient's former behavior and his current behavior. Pedophilia normally emerges in adolescence and normally emerges gradually, in step with all the other sexual changes we go through. But when a sixty-year-old man with a hitherto staid sex life—as happened in one case—starts having sex with his underage daughter and recklessly pursuing prepubescent boys, a neurologist should probably take a look. (This man also began sodomizing cattle out of the blue and adorning his penis with red ribbons.) Even this criterion doesn't cover every case, though. It wasn't a crime, but S.M., the woman with amygdala damage, repeatedly propositioned her doctors for sex: in addition to losing all fear of snakes and muggers, she'd apparently lost all social fear as well. Yet her condition emerged slowly, over many years.

These cases not only raise difficult questions about culpability, they introduce quandaries about how to punish offenders. If brain damage caused the criminal behavior, you might be tempted to be lenient, since on some level it's not someone's fault. But some judges (and scientists) reason the exact opposite way: if someone has permanent brain damage that leaves him with horrendous impulse control and an appetite for young girls, rehabilitation might do no good. Perhaps better to toss him into a supermax.

There's no doubt that neuroscience will change our judicial system, but no one quite knows how. Neuroscience can help us understand why someone like Harvey Cushing periodically erupted—why his anger overwhelmed his decency when he noticed that an assistant had missed the left parathyroid body. It can help us understand why S.M. lacked fear, or why a man might find a safety pin sexy. But if a brain-damaged person attacks someone because his frontal lobes can't control a cascade of emotions, then even if we can trace the causes down to the last neuron, neuroscience alone cannot tell us what to do next. Determining that will require a lot of hard thought and careful reasoning—which means that it will require listening to our emotions, to supplement our reasoning and make it more humane. If emotion without reason is blind, it's equally true that reason without emotion is lame: a world run by Elliots would be a disaster. So despite all the advances of neuroscience, all the fancy machines and illuminating insights, we still need our old, wet gray matter—the only place where emotion and reason come together and alchemize into what we call wisdom—to tell us how to act.

PART IV

BELIEFS AND DELUSIONS

CHAPTER EIGHT

The Sacred Disease

In this section we shift from the physical brain to the mental brain. Common sense tells us there's a sharp distinction between the physical and the mental, but diseases like epilepsy show how fuzzy the boundaries are.

Neurosurgeon Wilder Penfield had waited days for the letter about his sister, and when it arrived, he felt stupid, so stupid. A telegram a few days before had said little, only enough to distress him: that his sister Ruth was ailing, and that she and their mother had boarded a train from Los Angeles to Montreal, to seek his professional opinion. The letter that arrived on December 1, 1928, explained more. It said that Ruth, then forty-three, had suffered an increasing number of seizures in the past decade. These had included one two-day barrage of fits and one giant convulsion that required CPR to revive her. Now fits shook her almost daily, and she seemed likely to die without treatment.

Neurosurgeon Wilder Penfield. (National Library of Medicine)

Reading this, Penfield's mind flashed back to an ugly incident from their childhood in Wisconsin. He, fourteen, stood eavesdropping outside Ruth's bedroom door; she, twenty, lay prostrate inside, rigid and immobile, her head and neck spasming and jerking. He couldn't have known her diagnosis then, but by 1928 he'd become a world expert on epilepsy. And yet until he received the letter, Penfield had never put it all together, never realized that all Ruth's "headaches" and "nervous spells" over the years had been seizures. *How did I miss it?* Their prim Presbyterian family never discussed illness, and for the past decade he'd been too busy to inquire much into Ruth's health. Now he would have to confront her epilepsy directly: she was due to arrive in Montreal in a few hours.

As a surgeon Penfield was unorthodox. He stood out for his willingness to cut the brain—to scoop out whole handfuls of tissue to make sure he removed every last diseased cell. "No brain is better than bad brain," he once said. Paradoxically, though, he also viewed the human brain with something like reverence. For he believed that, hidden deep within the brain, lay the seat of human consciousness—the wellspring of our inner minds, our inner selves, our inner (he wasn't afraid to say it) souls.

Penfield's desire to glimpse the inner essence of human beings had pushed him into neurosurgery in the first place. He'd been something of an All-American puke at Princeton University—class president, starting football tackle, a meathead who outgrew his shirts because his neck got too wide. He hobnobbed with former Princeton president Woodrow Wilson, and wealthy alums feted Penfield and his teammates at the Waldorf-Astoria. But after the Saturday gridiron glory, Penfield spent most Sundays teaching Sunday school, and he had considered joining the ministry before deciding it wasn't manly enough.

Medicine had initially turned him off, mostly because of its association with his father, a philandering physician who'd abandoned the family to roam the woods and live off the land. But a medical-minded friend convinced Penfield to help him bluff their way into an emergency room in New York City one afternoon during college. Posing as a resident, Penfield watched four different operations and became fascinated with surgery. He even began shaving with a Sweeney Todd–style straightedge to steady his hands. So when this Big Man on Campus won a Rhodes Scholarship in 1914, he decided to study physiology at Oxford and prep for medical school.

At Oxford he encountered mostly other Americans (including a sour, unknown poet named T. S. Eliot) because the British lads were off shivering in muddy trenches or getting shot out of the sky over France. English children taunted Penfield as a shirker, so he volunteered at hospitals in France during his breaks, once again bluffing his way into treatment centers. (The first time he ever handled chloroform, he had to knock out a man for emergency surgery.) While crossing the English Channel during spring break in 1916, a German torpedo exploded right beneath where he stood on deck, flinging him twenty feet straight up. He landed dazed, with a mangled right knee, and just managed to crawl astern as the bow cracked off and sank. Rescuers plucked him off the bobbing wreckage, and he reached the hospital in such bad shape that a newspaper near his hometown ran his obituary.

During his convalescence Penfield decided that God must have spared him for some higher purpose. Over the next decade he determined that that purpose was to illuminate the age-old mind-body problem—how a material brain produces an immaterial mind. The question had first tickled him at Oxford, in research labs where scientists had removed the upper brains of cats. These cats ate and slept and moved about just fine, but they'd become zombies—any sense of playfulness or personality vanished. Extrapolating, Penfield wondered where humankind's highest faculties sat, and he decided to find out. For a junior surgeon to take on the mind-body problem when Aristotle, Descartes, Cajal, and other luminaries had failed wasn't hubris, or at least not only hubris. New neurosurgical techniques were finally allowing scientists to work directly with the living brain—to prod it, palpate it, probe it with electricity. The prospect thrilled Penfield, and he spent the next few decades trying to glimpse the "ghost in the machine."

Such were the high-minded thoughts that had distracted Penfield for years. The news about Ruth brought him crashing back down to the stark medical realities of life and death. He'd accepted the job in Montreal only months before, and she arrived at his new house looking dazed and groping for handholds. Even before she ate breakfast, Penfield sat her down and shined a light into her eyes. Her optic nerve looked swollen, and he spotted little red hemorrhages on her retina, like cracks in a failing dam. He knew immediately, and had to steady himself on her shoulder. A tumor behind her sinuses was crushing her brain.

Someone had to operate, soon. So after he sent Ruth to lie down, Penfield gathered three colleagues in his parlor and nominated himself. His aggressive approach would serve Ruth best, he argued: pussyfooting around, by leaving too much tissue behind, would only condemn her down the line. That said, Penfield knew that a sensible physician doesn't treat loved ones—seeing them laid bare can make even the steadiest hands tremble—so he asked his colleagues for advice. They debated a long while, but permitted him to scrub in.

On December 11, 1928, Ruth drank a high-protein shake for breakfast at the hospital. Nurses "Bic'd" her head down to the smooth scalp and sterilized her skin. Penfield then used a wax crayon to outline a horseshoe above her right eyebrow. He sawed around the horseshoe and flipped open a trapdoor of skull, exposing her brain. A nearby atomizer kept the surface glistening with puffs of saline.

He paused here and asked Ruth how she felt. Fine, she answered.

Because the brain's surface cannot feel pain, Ruth could stay awake for the operation, with only novocaine to dull her scalp, similar to what you'd get at the dentist. Penfield in fact preferred that patients remain conscious during surgery* and chat with the nurses, because that way he knew their brains were still functioning. (The danger started when they fell silent.) Not long into this operation, though, hearing Ruth chitchat about her six children unnerved him. Just this once, he asked his patient to hush.

After decades of growth, the tumor had consumed most of Ruth's right frontal lobe: it looked like a gray octopus sucking on her brain, with plenty of fat blood vessels feeding it. And although it consisted only of glial cells, its sheer bulk was crushing nearby neurons, which then misfired and caused seizures. Penfield set about removing the mass piece by piece; it felt somewhat hard to the touch, like crusty dough. In all, he had to remove one-eighth of her brain because of collateral tissue damage, the largest excavation he'd ever undertaken. From above, the remainder of her brain looked like a lopped-off loaf of focaccia.

And that wasn't the worst of it. As Penfield prepared to sew his sister back up, he noticed that one stray root of the tumor had slithered off along the skull floor. His eyes followed it into a hidden recess. The assisting surgeon noticed Penfield's interest and murmured, "Don't chance it." But Penfield had nominated himself for the surgery exactly because he was the type to chance it—why leave a malignancy behind? So in what he later called "a frenzy"—"I was rather reckless," he admitted to Ruth's husband—he decided to attack. He wrapped a loop of silk thread around the final arm, drew it shut like a noose, and tugged.

The arm pinched off and came free. Unfortunately, a nearby blood vessel also tore loose, and Ruth's skull flooded. Penfield snatched some cotton wads to stanch the flow and pressed down hard, but her brain was disappearing beneath a rising red sea. Many tense minutes passed, and Ruth lost consciousness; only after three blood transfusions did she stabilize. Just as he sopped up the last bits of blood, however, thinking he'd won his gamble, he saw that the tumor had burrowed still deeper. It actually extended into Ruth's left hemisphere, beyond where he could reach. At this, he wilted. The operation was over; the tumor had won. Years later he would revisit this moment—or rather, it would revisit him.

During the next few days Ruth suffered the expected post-op headaches and nausea, but her memory, sense of humor, and stamina returned quickly. Three weeks after going home, in February, she even sent Penfield a letter about a recent night out at the Rotary Club, dancing with her husband. She'd felt spry and sexy and just so alive in her blue hat and blue dress. She told her brother that he'd given her her life back.

Still, those close to Ruth noticed problems. Most important, she lacked what neuroscientists call an "executive sense," meaning that she now struggled to form plans and see them through. (Elliot, in our previous chapter, also suffered from this.) Penfield observed this deficit firsthand in early 1930 on a trip to California, when he visited Ruth for dinner one night, a simple affair for five. She'd had all day to prep, and it should have been patty-cake for an experienced homemaker like her. But Penfield arrived in the late afternoon to find her in tears—her children running wild, the table unset, the ingredients for the salad and sides strewn across the counters. The night didn't turn out badly after all: Ruth could still follow directions and cook just fine, so after Penfield calmed her down and got the roast roasting, she cheered up. Penfield, though, could only sigh: Ruth was no longer Ruth.

In the end, Penfield's surgery merely bought Ruth time—time she and her family cherished, but too little of it. In May 1930 the seizures returned, and her eyes began bulging again. Unable to face another surgery, Penfield sent her to Harvey Cushing in Boston. When Cushing opened Ruth up, he saw that the octopus had grown back, as ugly and greedy as before. Cushing, a more cautious surgeon than Penfield, scraped out what he could, but more seizures erupted six months later. At this point Ruth declined further treatment (she'd recently converted to Christian Science), and in July 1931 a stroke finally killed her.

Ruth's death brought Penfield back to the dark hours after her first surgery. After showering up that day, he'd slumped down on a bench in the surgeons' dressing room, wrapped in a towel and near tears. He was probably the world's most gifted young neurosurgeon, and he'd been whipped. But he'd recently learned a new surgical technique while on sabbatical in Germany. It involved stimulating the cortex with electricity to find the origins of people's seizures. From a surgical point of view, the idea seemed promising—a way to cut down on guesswork and pinpoint exactly what tissue to remove. But Penfield had the vision to see the technique's broader potential. Sparking the brain in different places often induced hallucinations, stray sounds, or muscle spasms in the patient—sensations unrelated to the seizures but interesting in their own right. At the time scientists had only just started to explore the brain's topography, and Penfield realized that electrical stimulation could help them map the cortex with far more precision. What's more, the technique might even help him solve the mind-body problem, since he could probe patients' minds while they were awake...

Penfield woke up from this reverie in the dressing room half naked, with one sock on his foot, the other still in his hand. He had no idea how long he'd been muttering to himself, perhaps for hours. But after he snapped to, he vowed to do something he'd long been pondering: establish a new neurological institute, so he could study the conscious brain in detail. Ruth's death reminded him of this promise and finally spurred him. The institute opened within a decade, and over the next twenty years Penfield probably did more than any other scientist to explain how the brain works in real time. And while he never did resolve the big metaphysical questions that God had spared him to answer, he did find something almost as amazing—fragments, traces, glimpses of what we might call the scientific equivalent of the soul.

~

For most of recorded history, human beings situated the mind—and by extension the soul—not within the brain but within the heart. When preparing mummies for the afterlife, for instance, ancient Egyptian priests* removed the heart in one piece and preserved it in a ceremonial jar; in contrast, they scraped out the brain through the nostrils with iron hooks, tossed it aside for animals, and filled the empty skull with sawdust or resin. (This wasn't a snarky commentary on their politicians, either—they considered everyone's brain useless.) Most Greek thinkers also elevated the heart to the body's summa. Aristotle pointed out that the heart had thick vessels to shunt messages around, whereas the brain had wispy, effete wires. The heart furthermore sat in the body's center, appropriate for a commander, while the brain sat in exile up top. The heart developed first in embryos, and it responded in sync with our emotions, pounding faster or slower, while the brain just sort of sat there. Ergo, the heart must house our highest faculties.

Meanwhile, though, some physicians had always had a different perspective on where the mind came from. They'd simply seen too many patients get beaned in the head and lose some higher faculty to think it all a coincidence. Doctors therefore began to promote a brain-centric view of human nature. And despite some heated debates over the centuries—especially about whether the brain had specialized regions or not—by the 1600s most learned men had enthroned the mind within the brain. A few brave scientists even began to search for that anatomical El Dorado: the exact seat of the soul within the brain.

One such explorer was Swedish philosopher Emanuel Swedenborg, one of the oddest ducks to ever waddle across the stage of history. Swedenborg's family had made a fortune in mining in the late 1600s, and although he was raised in a pious household—his father wrote hymns for his daily bread and later became a bishop—Swedenborg devoted his life to physics, astronomy, and geology. He was the first person to suggest that the solar system formed when a giant cloud of space dust collapsed in upon itself, and much like Leonardo he sketched out plans for airplanes, submarines, and machine guns in his diaries. Contemporaries called him "the Swedish Aristotle."

In the 1730s, just after turning forty, Swedenborg took up neuroanatomy. Instead of actually dissecting brains, though, he got himself a comfy armchair and began leafing through a mountain of books. Based solely on this inquiry, he developed some remarkably prescient ideas. His theory about the brain containing millions of small, independent bits connected by fibers anticipated the neuron doctrine; he correctly deduced that the corpus callosum allows the left and right hemispheres to communicate; and he determined that the pituitary gland serves as "a chymical laboratory." In each case Swedenborg claimed that he'd merely drawn some obvious conclusions from other people's research. In reality, he radically reinterpreted the neuroscience of the time, and most everyone he cited would have condemned him as a luna- and/or heretic.

The history of neuroscience might look quite different if Swedenborg had pursued these studies. But in 1743 he began to fall into mystical trances. Faces and angels hovered before him in visions, thunder pealed in his ears; he even smelled hallucinatory odors and felt odd tactile sensations. In the midst of these trances he often fell down shuddering, and an innkeeper in London once found him wrapped in a velvet nightgown, frothing at the mouth and babbling in Latin about being crucified to save the Jews. Swedenborg woke up insisting he'd touched God, and at different times claimed to have conversed with Jesus, Aristotle, Abraham, and inhabitants of the five other planets. (Uranus and Neptune hadn't been discovered yet, or he surely would have met Uranians and Neptunians, too.) Sometimes the visions revealed answers to scientific mysteries, such as how bodies eaten by worms will nevertheless be reconstituted on Judgment Day. Other trances were more casual, like the time he brunched with angels, and discovered that some angels hate butter. Yet another time God pulled a mean joke and turned Swedenborg's hair into a Medusa's nest of snakes. Compared to such intense visions the cerebral pleasures of science had no chance, and from 1744 onward he devoted his life to chronicling these revelations.

Swedenborg died in 1772, and history has returned a split verdict on his legacy. His eclectic dream diaries charmed the likes of Coleridge, Blake, Goethe, and Yeats. Kant, meanwhile, dismissed Swedenborg as "the arch-fanatic of all fanatics." Many other observers were similarly baffled. What could transform a gifted and reserved gentleman scientist into someone whom John Wesley called "one of the most ingenious, lively, entertaining madmen that ever set pen to paper"? The answer may be epilepsy.

At its most basic level epilepsy involves neurons firing when they shouldn't and stirring up storms of electrical activity inside the brain. Neurons can misfire for many reasons. Some misfit neurons were born with misshapen membrane channels and can't regulate the flow of ions in and out. Other times, when axons suffer damage, neurons start discharging spontaneously, like frayed electrical wires. Sometimes these disturbances remain local, and just one location in the brain goes on the fritz, a so-called partial seizure. In other cases, the seizure short-circuits the entire brain and leads to either a grand mal or a petit mal seizure. Grand mals (now called tonic-clonic seizures) start with muscular rigidity and end with the stereotypical thrashing and foaming; they're what most of us think of when we think of epilepsy. Petit mals avoid the thrashing but usually cause "absences," in which the victim freezes up and her mind goes blank for a spell. (William McKinley's wife, Ida, suffered from petit mals. During state dinners McKinley sometimes just draped a napkin over her face and blustered through the next few minutes to divert attention.)

The triggers for epileptic fits can be bizarrely specific: noxious perfume, flashing lights, mah-jongg tiles, Rubik's cubes, wind instruments, parasitic worms. Although potentially embarrassing, seizures don't always compromise someone's quality of life—and in rare cases, people benefit. Some first-time seizure victims find that they can suddenly draw much better or that they now appreciate poetry. Some folk (but only women so far—sorry, guys) orgasm during seizures. Specific triggers aside, seizures do erupt most commonly during times of stress or psychological turmoil. Probably the best example of this is Fyodor Dostoyevsky.

Biographers disagree about whether Dostoyevsky suffered any seizures when young, but he himself said that his epilepsy emerged only after his near execution in Siberia. Dostoyevsky and some fellow radicals were arrested in April 1849 on charges of plotting to overthrow Czar Nicholas. That December soldiers dragged the lot of them to a snowy public square studded with three tall posts. Until that moment the comrades assumed they'd get off with breaking rocks for a spell. Then a priest arrived, as did a firing squad, and clerks handed the prisoners white smocks to change into—funeral shrouds. Dostoyevsky grew frantic, especially when a friend pointed to a cart filled with what looked like coffins. Soldiers meanwhile marched the crew's ringleaders to the posts and covered their eyes with white hoods. The gunmen raised their rifles. A minute of agony passed. Suddenly the rifles dropped, and a messenger clattered up on horseback, carrying a pardon. In reality Nicholas had staged the entire scene to teach the punks a lesson, but the stress unhinged Dostoyevsky. And after he'd spent a few months in a labor camp (the czar didn't let them off *that* easy), the abusive guards and harsh weather finally pushed him over the edge, and he had his first major fit—shrieking, foaming, convulsions, the whole production.

That first seizure lowered the threshold inside Dostoyevsky's brain, and after that, any mild stressor, mental or physical, could fell him. Guzzling champagne could trigger fits, as could staying up all night to write or losing money at roulette. Even conversations could detonate him. During a philosophical bull session with a friend in 1863, Dostoyevsky began pacing back and forth, waving his arms and raving about some point. Suddenly he staggered. His face contorted and his pupils dilated, and when he opened his mouth a groan escaped: his chest muscles had contracted and forced the air out. The seizure that followed was intense. A similar incident occurred a few years later, when he collapsed onto the divan in his wife's family's living room and began howling. (This couldn't have impressed the in-laws.) Dreams could set him off as well, after which he usually wet the bed. Dostoyevsky compared the seizures to demonic possession, and he often plumbed the agony of them in his writing, including epileptic characters in *The Brothers Karamazov, The Insulted and Injured,* and *The Idiot.*

Dostoyevsky almost certainly had temporal lobe epilepsy. (As mentioned, the temporal lobes sit behind your temples and wrap laterally around the brain, somewhat like earmuffs.) Not all temporal lobe epileptics thrash and foam, but many of them do experience a distinctive aura. Auras are sights, sounds, smells, or tingles that appear during the onset of seizures—a portent of worse things to come. Most epileptics experience auras of some sort, and most non–temporal lobe epileptics find them unpleasant: some unlucky folk smell burning feces, feel ants crawling beneath their skin, or pass horrendous gas. But for some reason—perhaps because the nearby limbic structures get revved up—auras that originate in the temporal lobes feel emotionally richer and often supernaturally charged. Some victims even feel their "souls" uniting with the godhead. (No wonder ancient doctors called epilepsy the sacred disease.) For his part, Dostoyevsky's seizures were preceded by a rare "ecstatic aura" in which he felt a bliss so intense it ached. As he told a friend, "Such joy would be inconceivable in ordinary life...complete harmony in myself and in the whole world." Afterward he felt shattered: bruised, depressed, haunted by thoughts of evil and guilt (familiar motifs in his fiction). But Dostoyevsky insisted the hardship was worth it: "For a few seconds of such bliss I would give ten or more years of my life, even my whole life."

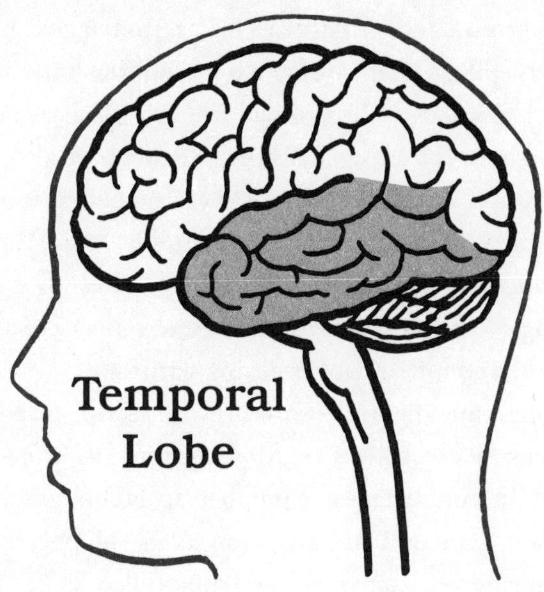

Temporal lobe epilepsy has transformed other people's lives in a similar way. All human beings seem to have mental circuits that recognize certain things as sacred and predispose us to feeling a little spiritual. It's just a feature of our brains (Richard Dawkins excepted, perhaps). But temporal lobe seizures seem to hypercharge these circuits, and they often leave victims intensely religious, as if God has personally tapped them as witnesses. Even if victims don't become religious, their personalities often change in predictable ways. They become preoccupied with morality, often losing their sense of humor entirely. (Laugh lines are few and far between in Dostoyevsky.) They become "sticky" and "adhesive" in conversations, refusing to break them off despite pretty strong signs of boredom from the other party. And for whatever reason, many victims start writing compulsively. They might churn out page after page of doggerel or aphorisms, or even copy out song lyrics or food labels. The ones who visit heaven often chronicle their visions in excruciating detail.

Based on these symptoms, especially the rectitude and sudden spiritual awakening, modern doctors have retrodiagnosed certain religious icons as epileptics, including Saint Paul (the blinding light, the stupor near Damascus), Muhammad (the trips to heaven), and Joan of Arc (the visions, the sense of destiny). Swedenborg also fits the profile. He converted abruptly, he wrote like a methamphetamine addict (one book, *Arcana Coelestia,* runs to two million words), and he often shuddered and fell down senseless during visions. On occasion he even felt "angels" thrusting his tongue between his teeth as if to make him bite it off, a common danger during seizures.

At the same time there are problems with casting Swedenborg and other religious folk as epileptics. Most seizures last a few seconds or minutes, not the hours that some prophets spend immersed in trances. And because a temporal fit can paralyze the hippocampus, which helps form memories, many temporal lobe epileptics can't remember their visions in much detail afterward. (Even Dostoyevsky lapsed into vague descriptions when recounting their actual content.) Also, while Swedenborg's trances in particular blended sights, sounds, and smells into a heady, heavenly froth, most epileptics hallucinate with one sense only. Most damningly, most epileptic auras are tedious, producing the same refulgent light, the same chorus of voices, or the same ambrosial smells time and again.

So while epilepsy might well have induced their visions—the idea makes sense—it's important to remember that Joan of Arc, Swedenborg, Saint Paul, and others also transcended their epilepsy. Probably no one but Joan would have rallied France, no one but Swedenborg would have imagined angels eating butter. As with any neurological tic, temporal lobe epilepsy doesn't wipe someone's mental slate clean. It simply molds and reshapes what's already there.

~

Research on electrical activity within the brain, including seizures, has done more than merely shed light on the origins of religious sentiment. It also illuminated one of the eternal debates in the history of neuroscience: whether the brain has specialized parts that control different mental faculties, or whether, like the indivisible soul, the brain cannot be subdivided into smaller units.

Indivisiblists held sway through the middle of the nineteenth century, but the worm started to turn in the 1860s. Paul Broca discovered in 1861 that many people who'd lost the power to speak had lesions in the same part of the frontal lobe. (More on Broca and language later.) Around that same time, English neurologist John Hughlings Jackson noticed that many epileptics had uncannily similar seizures. These weren't grand mal eruptions or temporal lobe raptures, but mild shaking palsies that started in one spot and "marched" up and down the body in unvarying order. If the big toe started trembling, then the foot, calf, and leg always followed, in that order. If the elbow started quaking, then the forearm, hand, and individual fingers followed. Jackson deduced that the brain must contain a body map with discrete territories, and that seizure hurricanes must be roaming across this map from region to region. This research had a special poignancy for Jackson, because one of the epileptics he studied was his wife, Elizabeth, who died at forty of complications from the disease. Although never a warm man—Jackson rarely bothered to remember patients' names—Elizabeth's death devastated him, and he became a semirecluse.

Localization research got another boost in the early 1870s. First, a pair of bearded Berliners, Gustav Fritsch and Eduard Hitzig, began a series of experiments on the brains of anesthetized dogs. They ran most of these experiments in Hitzig's spare bedroom, strapping the dogs down on Frau Hitzig's dressing table. By sparking different spots in the brains, the duo managed to flail the dogs' limbs and twitch their faces. Another scientist one-upped them in 1873: he made cats extend their paws as if playing with twine, made dogs retract their lips as if snarling, even made a rabbit dismount from a table with a backward somersault. Both sets of experiments proved that electricity could excite the brain's surface, and they provided a crude map of movement and sensation centers.

However tantalizing, this work didn't impress everyone, mostly because it involved lower animals. No doubt the human brain differed, perhaps significantly. To confirm the existence of specialized brain regions in human beings, then, scientists needed a human test case. She finally emerged in 1874, in Ohio. Her story could have been a triumph of nineteenth-century medicine. Instead it became a prime example of scientific hubris and abuse of power.

After serving in the Union army, an impressively bearded doctor named Roberts Bartholow moved to Cincinnati in 1864. Although considered icy, he attracted flocks of patients, and he soon opened up one of the country's first "electrotherapeutic rooms" in the Good Samaritan Hospital. The room included a chair for patients to sit in, as well as a few generators: one produced alternating current, and looked like an oversized sewing machine with metal coils around it; the other produced direct current, and looked like a wooden pantry stuffed with fluid-filled mason jars. The electricity from the devices flowed into metal suction cups or slender metal probes, which Bartholow used to treat polyps, cancer, hemorrhoids, paralysis, impotence, and pretty much every other ailment. He even crafted special sponge "slippers" to tingle his patients' feet.

The experiments on animal brains had enraptured Bartholow, and some historians suspect that as soon as poor Mary Rafferty removed her wig in his office, the forty-two-year-old doctor made up his mind what to do. Rafferty, a feebleminded thirty-year-old Irish maid, had fallen into a fire one morning as a lass and had burned her scalp so badly that the hair never grew back. She wore a wig over her scars, but in December 1872 a malignant ulcer opened up beneath it. Rafferty blamed the ulcer on the wig's sharp whalebone frame, which dug into her skin; Bartholow diagnosed cancer. Whatever the cause, when Rafferty arrived at Good Samaritan in January 1874, a two-inch hole had opened in her skull, and a wide-eyed Bartholow could see her parietal lobes pulsating.

Doing what they could, the nuns who served as hospital nurses

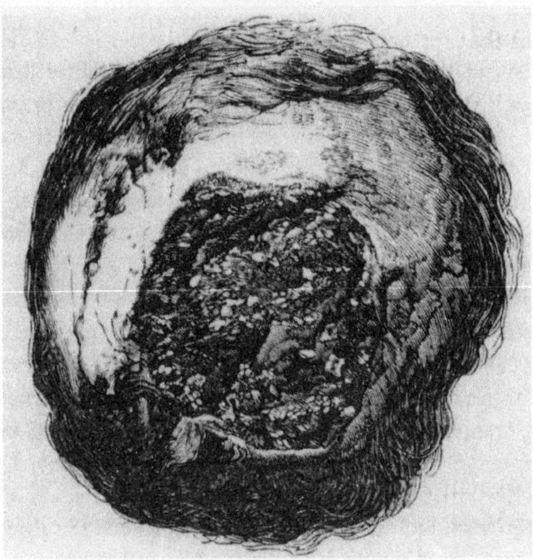

The exposed brain of Mary Rafferty, the subject of one of the most unethical experiments in medical history.

dressed and re-dressed Rafferty's wound. But she just didn't get any better, and by March she was clearly dying. Around this time Bartholow sidled up to Rafferty and, oozing charm, asked her about undergoing some tests. In defending himself later, Bartholow recalled that Rafferty had "cheerfully" agreed. Given her feeblemindedness, she likely didn't grasp what she'd agreed to. Bartholow sat her down in the electrotherapeutic room anyway and unwrapped her turban of bandages. He then slid two needle-shaped electrodes into her gray matter and closed the switch on the sewing-machine generator.

Based on Rafferty's reaction, Bartholow must have tapped into her motor centers: her legs kicked out, her arms flailed, her neck twisted backward like an owl's. Bartholow later claimed that she'd smiled during this macabre dance, but given that she also shrieked throughout, her facial muscles had probably just twisted into a rigid mockery of mirth. (The brain's surface cannot feel pain, but its lower regions can; sparking the brain can also cause pain in the body.) Because she kept smiling Bartholow pushed on, moving the electrified needles around and increasing the current for "a more decided reaction." He got it. Her pupils dilated, her lips turned blue, her mouth frothed. She began breathing erratically and soon suffered a seizure, five full minutes of thrashing. Bartholow called it a day at that point, and Rafferty collapsed into bed pale and dizzy. Her pupils were also "blown"—dead and unresponsive. Bartholow nevertheless decided to tase her brain again a few days later, this time using the wooden cabinet generator. Understandably, the sight of the equipment induced a sort of posttraumatic seizure in Rafferty, and she fell unconscious ("stupid and incoherent," in Bartholow's words). Bartholow reluctantly postponed his experiments—and Rafferty died before he could resume them. An autopsy found needle tracks an inch deep inside her brain.

When Bartholow blithely published these results, the medical world turned on him like an autoimmune reaction: doctors worldwide howled, and the American Medical Association censured him. Chagrined but defiant, Bartholow countered that Rafferty had given informed consent—she'd said yes. And in spite of all the pious protestations, he argued, he'd proved what he'd set out to: the human brain had specialized regions of function, which scientists could probe with electricity. Having staked his claim as a pioneer, Bartholow did admit that, given the suboptimal outcome (i.e., death), *repeating* the experiment "would be in the highest degree criminal." But how could he have known beforehand? This half apology absolved Bartholow in some circles, and his career never really suffered: he built the largest practice in Cincinnati, cofounded the American Neurological Association, and earned honorary degrees in Edinburgh and Paris. But the fiasco probably did retard the study of the living human brain, since other scientists didn't want the stink of another Mary Rafferty on them.

Although a few scientists (like Harvey Cushing) did probe the living brain with electricity during the next few decades, the work proceeded fitfully, and it took a man of Wilder Penfield's stature to fully rehabilitate the field. Penfield's career did not get off to the most promising start, actually: his first two surgical patients died, a common occurrence in the 1920s.* Nevertheless, Penfield honed his techniques and by the end of the 1920s was taking on the most difficult cases of epilepsy out there. Many epileptics had scars or tumors inside their brains, and in those cases the operation was as straightforward as neurosurgery can ever be: just scoop out the offending tissue. Penfield, though, also took on patients without obvious trauma or damage—a much trickier procedure, since it wasn't clear where the epicenter of the seizure was.

To find the epicenter, Penfield became, essentially, a cartographer. Because so few people had explored a conscious brain before, whole continents of the neural hemispheres remained as sketchy as early-1500s maps of the Americas. So Penfield decided to create a better map, by using electricity as his compass and sextant. The work really got going in 1934, when the institute that he'd vowed to establish after Ruth's death finally opened in Montreal, a $1.2 million affair (roughly $21 million today) nicknamed the Neuro. The Neuro attracted scores of brilliant scientists—David Hubel of the cat-vision experiments got his start there—but it was Penfield's mapping work that proved the most influential.

At least superficially this work resembled Roberts Bartholow's experiments on Mary Rafferty, in that Penfield was using electrified wires to spark the brain's surface. Penfield, however, used lower and more precise voltages. And rather than treat the patient as a passive tool—zapping her brain and seeing what the hell happened—Penfield collaborated with each patient, gently stimulating various spots on her cortex and asking what she felt at each point.

Often she felt nothing. But when she did experience something, Penfield dropped a marker—a numbered piece of confetti—onto that square millimeter of tissue, and a secretary behind a glass wall recorded the result. The kinds of reactions varied geographically around the brain. If Penfield stimulated the visual cortex (in the back), the patient might see lines, shadows, or crosses—the constituent elements of sight. If he stimulated the auditory cortex (above the ears) she might hear ringing, hissing, or thumping. If he stimulated the movement and tactile centers she might begin swallowing violently or comment, "Mah thongue theems tah be pallalyzed." More provocatively, stimulating the speech centers often forced the patient to sing against her will—an aria of *aaaaah* that got louder every second. Penfield did have a sense of mischief and sometimes got patients talking only to cut them off: "I visited my daughter yesterd-*aaaaah*—" He challenged another man to keep quiet no matter what—to try as hard as he could not to say anything. The patient bit down, and Penfield even warned him when the zap was coming. It didn't matter: the man sang like a canary. "I win," Penfield said. The man laughed.

These neurological soundings improved brain surgery in two ways. First, Penfield often succeeded in triggering a patient's aura at some point. This wasn't always a pleasant process, since the auras might include nausea, dizziness, or foul smells. But when he'd zeroed in on that sensation, he knew what folds of tissue to remove in order to interrupt the seizure circuit. Second, and just as important, Penfield knew what *not* to remove. He always began his surgeries by mapping out the boundaries of the patient's movement and speech centers. He could then steer clear of those centers when excavating tissue.

Determining what areas to avoid had an unexpected side benefit, in that it allowed him to map the brain's movement and tactile centers in unprecedented detail. No one before Penfield knew that the face region lies next to the hand region, or that the face, lips, and hands all owned huge, Canada-sized territories. These discoveries laid the groundwork for understanding phantom limbs in later decades. In a broader sense they also demonstrated just how unusual the brain's view of the body is. To drive that point home, Penfield sketched a famous cartoon, the "cortical homunculus," in the 1950s, a vision of what human beings would look like if the size of each body part corresponded to the amount of cortical territory devoted to running it. Turns out we'd all have Q-tip legs, bee-stung lips, and huge mitts for hands—within our brains we all look like bad Giacomettis.

A sensory homunculus, after Wilder Penfield. The sensory homunculus and motor homunculus (not shown) are representations of what the body would look like if the size of each body part was proportional to the amount of gray matter dedicated to running it.

Penfield found evidence of brain rewiring as well. In truth, the atlas Penfield developed for the human brain was idealized—a platonic form that no individual brain conformed to. A language node in Adam, for instance, might sit several centimeters higher or lower than in Bob. And even within Adam, the language node might shift around year by year as his brain rewired itself, something Penfield noticed when he had to operate several times. Contrary to most scientists' expectations, then, each brain, each mind, had a unique geography. And its geography changed over time, since brain territories drifted like continental plates.

Of all the things Penfield uncovered about the brain, he cherished one discovery above all. It involved the temporal lobes, and he cherished it because it rose above the grubby, animalistic realms of touch and movement and sight, and soared toward the human soul. Neuroscientists had long neglected the temporal lobes, so when Penfield zapped the temporal lobe of a female patient in 1931 he had low hopes of finding anything good there. Instead of a typical sensation, though—a vague buzz, a patch of green light—her mind was transported back to the birth of her daughter twenty years before, an unusually crisp and specific vision. Penfield, bemused, never followed up on this. (He remembered thinking at the time, "It was never intended that men should understand [women] completely.") But five years later he provoked a similarly sharp memory in a teenage girl's temporal lobe. She was transported to an idyllic afternoon during her childhood, which she'd spent frolicking with her brothers in a field. Unfortunately a pervert had ruined everything by sneaking up behind her with a writhing burlap sack and asking, "How would you like to get into this bag with the snakes?" This memory happened to be the girl's seizure aura, so Penfield knew he had to excise this tissue. But this time Penfield took careful notes first, and afterward decided to investigate the temporal lobe further.

In fact, although he kept this work somewhat secret, Penfield spent the next two decades investigating as many temporal lobe visions as possible. Some people's visions proved mundane. One man saw a 7UP billboard. One woman pictured her dipsomaniac neighbor, Mr. Meerburger. Another woman heard an orchestra fading in and out every time Penfield lowered and raised his electrified wire, as if he were dropping a needle on a gramophone. (The woman in fact accused Penfield of hiding a phonograph in the operating theater.) Other visions, though, were more profound. People saw glimpses of heaven or heard angelic choruses—the sorts of auras that turn people toward religion. Several people saw their lives flash before their eyes, and one man hollered, "Oh, oh, God—I'm leaving my body!" and found himself hovering above his own operation.

At first, overly excited, Penfield thought he'd found the seat of human consciousness in the temporal lobes. He later revised that opinion and situated consciousness deeper down, somewhere near the brainstem. (This would explain why patients never lost consciousness during operations, even when surgeons were scooping out whole handfuls of upper brain. Later, however, we'll see why Penfield was wrong in his assumptions and why it makes little sense to search for a single locus of consciousness at all.) Regardless, Penfield maintained that working with the temporal lobes at least provided *access* to people's consciousnesses—a way to tap into their inner essences, perhaps even their innermost souls.

Such musings put Penfield outside the mainstream of neuroscience, but not too far outside. Historically, thinkers have always compared the brain to the technological marvels of the age: Roman doctors likened it to the aqueducts; Descartes saw a cathedral organ; scientists during the Industrial Revolution spoke of mills, looms, clocks; in the early 1900s the telephone switchboard was in vogue. Those are materialistic analogies, but neuroscience has always tolerated a fair amount of mysticism as well. Andreas Vesalius's *Fabrica* provoked so much hatred in part because its precise renderings of the brain left no fuzzy places where the soul might camp out. Later generations of neuroscientists had even stronger spiritual bents. For his part, Penfield tried to split the difference: he compared the human brain to a computer, but insisted that the brain had a programmer as well—an immaterial essence that operated it.

Still, there's no denying that neuroscientists have gotten more materialistic over the past century: the old saying that "the brain secretes thought the way the liver secretes bile" pretty much sums up their metaphysics. Penfield's religious convictions, however, only deepened as he aged, especially as he found new spiritual outlets: at age fifty, for instance, he began laboring over a religious novel, a bildungsroman about Abraham entitled *No Other Gods.* Like Silas Weir Mitchell, he found he couldn't get at some truths about the human condition except through stories. (Penfield later published a second novel, about Hippocrates, who'd studied epilepsy and the mind-body problem in ancient Greece.)

Penfield even dared to lecture here and there about how the mind arises from the brain, lectures in which he quoted Job and Proverbs and slyly promoted mind-body dualism. He got away with this because he'd built his dualism on a lifetime of surgical observations. For example, even though he could make his patients kick or bleat during surgery, he emphasized that the patients always felt they'd been forced to act. He never succeeded in activating their *will* to act, which proved to him that the will lay beyond the material brain. Penfield also declared that mere electricity, while able to conjure up full-blown mental scenes, could never provoke real, high-level thinking: people heard orchestras playing but never composed music themselves, or got insight into mathematical theorems. Penfield saw real thinking as something you could never trick a brain into doing because, again, the mind lay somehow beyond the brain.

However tantalizing these ideas were, Penfield could never quite alchemize them into a coherent philosophy of mind, brain, and soul. So, just before turning seventy he retired from the bustle of surgery to pursue this work full-time. He wavered month to month between optimism and despair over how much headway he'd made into the mind-body-soul problem. He never lost his faith that the soul existed, nor that some people, like temporal lobe epileptics, communed directly with God. But Penfield convinced very few colleagues to take dualism seriously, and a flippant remark he'd made when young must have haunted his later years: "When a scientist turns to philosophy," he'd sneered, "we know he's over the hill."

Like Descartes and Swedenborg and so many others, Penfield never resolved the mind-body-soul paradox, and his evidence for dualism looks shakier every year. Among many other things, neuroscientists now know of brain patches that, when sparked, can indeed induce a desire to move or speak. It seems that free will is just another brain circuit. (More on this next chapter.) And while neuroscientists may not know exactly how the electrified tapioca inside our skulls gives rise to the glorious human mind, Penfield's solution—to decide ahead of time that we have a soul, and that the soul explains everything we don't understand about the brain—seems like a cop-out, a betrayal of the scientific ethos.

Nevertheless, unlike the vast majority of people who've bloviated about brains and souls, Penfield did make real, seminal contributions to neuroscience. "Brain surgery is a terrible profession," he once wrote to his mother. "If I did not feel it would become very different in my lifetime, I should hate it." Brain surgery did improve, not only during Penfield's life, but because of Penfield's life. And his innovative, unflinching approach to mapping the brain provided the first real glimpses of the ghost in the machine: the sensations and emotions—and even outright delusions—that make us human in the end.

CHAPTER NINE

"Sleights of Mind"

We've seen how emotions and other mental phenomena help us make decisions and form beliefs. But if those processes go awry — and they will — we fall into delusion.

To ensure everlasting peace on earth, Woodrow Wilson first had to conquer the U.S. Senate. After World War I ended, Wilson warned that civilization could not endure a second. He therefore wanted Congress to adopt the League of Nations treaty, which he saw as humanity's last, best hope for peace. But he met realpolitik opposition in the Senate, whose members felt the treaty would sacrifice national autonomy. So in autumn 1919 President Wilson took his case to the American people, embarking on an eight-thousand-mile, twenty-two-day speaking tour to stir up anger and break his opponents. The tour broke Wilson instead.

After stop number one, Seattle, Wilson and his entourage chugged down the Pacific coast, then circled east toward the Rockies. Already feeling weak, Wilson was clobbered by altitude sickness near Denver, and he stumbled getting onstage in Pueblo on September 25, thanks to a piercing headache. He nevertheless boarded a train for Wichita that afternoon. Just twenty miles down the track he fell ill, and his doctor suggested they halt the train and take a walk along a dirt road. On the stroll Wilson met a farmer who gave him a cabbage and apples, then hopped a fence to chat up a wounded private on a porch. He returned to the train refreshed. But at 2 a.m. he knocked on the door of his wife, Edith's, sleeping car, complaining of another staggering headache. More ominously, Wilson's physician, Cary Grayson, noticed that half, and only half, of the president's face had started twitching.

Grayson was already treating Wilson for various ailments—high blood pressure, intermittent migraines, intestinal distress (which Wilson referred to as "turmoil in Central America"). In retrospect, Wilson had probably also suffered two miniature strokes, in 1896 and 1906; Silas Weir Mitchell himself had examined the president-elect in 1912 and declared that Wilson wouldn't survive his first term. Year by year after that, Grayson had watched Wilson grow increasingly brittle. He'd even begged Wilson to skip the speaking tour in 1919—a proposal that Wilson had raged at, considering it insubordination. Now, outside Wichita, Grayson ordered the train to stop and suggested that Wilson cancel his remaining speeches. Uncharacteristically, the president relented, too weak to fight. Wilson spent much of the thirty-six-hour ride home staring out the window and occasionally weeping, the left half of his face sagging lower every hour.

Back in Washington a cruel cranial pain prevented Wilson from working, and he spent his days playing pool, riding in touring cars, and watching silent films in the White House cinema. Meanwhile, the League of Nations treaty stalled in the Senate. Wilson's nemesis, Senator Henry Cabot Lodge, even started mocking the literary quality of the League's grandiloquent charter, a charter penned by Wilson himself.

At 8:30 a.m. on October 2, Edith checked in on Wilson and found him awake in bed, weak and complaining of numbness. She made a crutch of herself beneath his shoulder, dragged him to the toilet, and left to summon Grayson. She returned to find Wilson crumpled on the floor, half naked and unconscious. She and Grayson immediately closed off Wilson's bedroom to all visitors, but a White House usher peeked in later and saw Wilson laid out like a waxwork, looking entirely dead except for the raw red gashes on his nose and temple, where he'd struck the exposed bathtub plumbing after collapsing.

For the next few months servants had to prop Wilson in a wheelchair each morning and hand-feed him. This latest stroke had paralyzed his left side, and he spent most days listening to Edith read or zoning out in the garden. Washington meanwhile lumbered along without him, because very few people knew about the stroke at first—it certainly wasn't public knowledge. In March 1915 Grayson had introduced Edith to the recently widowed Wilson, and Edith quid pro quoed by insisting that Wilson promote Grayson, a navy man, to rear admiral over dozens of better-qualified candidates. Now the two old chums conspired to conceal Wilson's health from most of his cabinet members and even the vice president—a constitutionally dicey situation. Before 1919 five presidents had died in office, most of them quickly; only Garfield had lingered, and he'd remained lucid. Not Wilson. In late November a press secretary painted a haunting image of Wilson as "a broken, ruined old man, shuffling along, his left arm inert, the fingers drawn up like a claw, the left side of his face sagging frightfully. His voice is not human; it gurgles in his throat, sounds like that of an automaton." In the vacuum of power* Edith essentially became the first female president, controlling what papers Wilson saw and sending out memos in his name but her handwriting.

Wilson resumed his duties as president after a few months, but he continued to struggle. He limped along with a cane, and photographers avoided shooting the melted left half of his face. Neurologically, he was, if anything, worse. Already an icy man, he grew even colder and more imperious, a sign of mental inflexibility. At the same time he was prone to weeping, a sign of emotional instability. Strangest of all, he stopped noticing things to his left. This wasn't eye trouble, since Wilson could, technically, see things on his port side: he didn't bump into furniture over there, for instance, since his unconscious brain could steer him around it. But he didn't consciously *notice* things on his left side unless someone pointed them out. As a hypothetical example, a dozen fountain pens could be piled on the left-hand side of his desk, but unless there was also one on the right-hand side, he'd complain about not having a pen at hand—as if the whole left side of his world didn't count. Bewildered aides had to rearrange his office, and they learned to usher guests in on his right side or else he'd snub them.

Eventually Wilson's intransigence doomed the League. He rebuffed all suggestions that he amend its charter—it was peace the Wilson way, or go to hell—and the ratification movement languished. Convinced he could nevertheless ram the League through Congress later, Wilson began campaigning for a third presidential term in 1920, even though he'd become a veritable recluse. Edith, Grayson, and others mercifully sabotaged this campaign at the Democratic National Convention in San Francisco that year by spreading rumors—actually the truth—about Wilson's impairments. The next year Wilson left the White House in tears, and even in his dotage remained convinced that he'd lost not one ounce of mental vigor. As late as January 1924 he sat down at his desk and sketched out a third inaugural address. He died within two weeks, his damaged brain finally exhausting itself.

Fifty years later, in a separate but equal branch of government, a Supreme Court justice reprised Wilson's sad farce. By 1974 William O. Douglas had become the longest-serving justice in Court history, having been appointed by FDR in 1939. He'd also become a globe-trotting liberal rabble-rouser and a pariah to conservatives: while Speaker of the House, Gerald Ford had tried to impeach him. On December 31, 1974, Douglas touched down in the Bahamas to celebrate the New Year there, accompanied by his fourth wife, a thirty-one-year-old blonde not even born when he'd joined the Court. Within hours of landing, Douglas had a stroke and collapsed.

Airlifted out of Nassau, he arrived at Walter Reed Hospital in Washington and spent the next few months convalescing. In all he missed twenty-one Supreme Court votes, and although his doctors noticed little progress—Douglas couldn't walk and his left side remained paralyzed—he refused to resign from the Court. Finally, in March, he badgered one doctor into granting him an overnight pass to visit his wife. Instead of heading home, though, Douglas instructed his driver (he certainly couldn't drive himself) to visit his office. He began catching up that very night and never returned to the hospital.

Douglas had some good reasons not to resign. His old enemy Gerald Ford had become president, and Douglas feared that Ford would "appoint some bastard" to take his place. Plus, the court would hear important cases that term on campaign finance and the death penalty. But mostly Douglas refused to resign because in his own mind there was nothing wrong. At first he told reporters that, far from having had a stroke, he'd merely stumbled, Gerry Ford–like, and gotten a little banged up—never mind the slurred speech and the wheelchair. When questioned about this, he claimed that the stories circulating about his paralysis were rumors, and he challenged naysayers to go hiking with him. When pressed further, he swore that he'd been kicking forty-yard field goals with his paralyzed leg that very morning. Hell, he said, his doctor wanted him to try out for the Redskins.

If anything, Douglas's performance away from reporters was more pathetic still. He started sleeping through hearings, forgetting names, confusing facts on important cases, and whispering to aides about assassins; because of chronic incontinence, his secretary had to drench his wheelchair in Lysol. The other eight justices, although bound by *omertà* not to pressure Douglas publicly, agreed to table all 4–4 ties until the next term and not let Douglas cast deciding votes. In a small concession to reality Douglas did seek specialized stroke treatment in New York during the summer recess in 1975, but he failed to improve. The other justices finally forced Douglas to resign in November—and even then Douglas kept returning to work, calling himself the "Tenth Justice," commandeering clerks, and attempting to cast more votes. *Nothing's wrong with me,* he insisted. It was a bewildering end for an eminent jurist.

Cases like Wilson's and Douglas's—both the result of parietal lobe damage—are depressingly familiar to neurologists. Wilson had

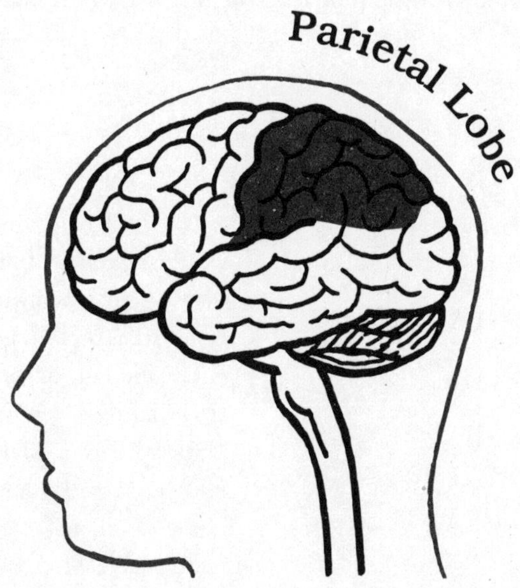

hemispatial neglect, the inability to notice half the world. Neglect patients shave only half their faces and dress only half their bodies. Ask them to copy a simple line drawing of a flower, and they'll bisect the daisy. Give them a bowl of salad, and they'll eat just half of it. Rotate the salad ninety degrees, and they'll eat just half the remainder. Their memories are bisected, too. Italian neuroscientists once asked a few hemispatial neglect victims to imagine standing in the famous square in their hometown, Milan, while facing the cathedral. When asked to name every building around the square, victims could remember the structures only on one side. The scientists then asked them to turn around mentally and face the opposite direction. At this point they could name all the buildings on the other side—but not a single one of those they'd named seconds before. Even blatant contradictions of logic and sense cannot penetrate hemispatial neglect. One man who neglected everything to his left got confused about how to draw 11:10 on a clock face. He finally drew the numerals six through twelve going up the wrong side—forcing the clock to tick counterclockwise. The paradox didn't distress him, though. Unlike regular stroke victims, who often lament their disabilities and grow depressed, people with hemispatial neglect usually remain blithe and cheerful.

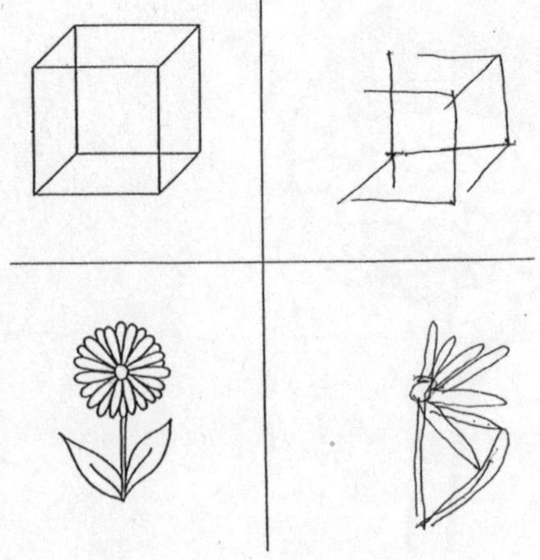

On the right half of the quadrant, drawings by a man who suffers from hemispatial neglect; he cannot notice anything to his left. (Masud Husain, from "Hemispatial Neglect," Parton, Malhotra, and Husain, *Journal of Neurology, Neurosurgery, and Psychiatry* 75, no. 1 [2004]: 13–21)

Along those same lines, some stroke victims, like Douglas, refuse to admit they're paralyzed, and they lie shamelessly to themselves and others to maintain a delusion of competence. (Doctors call these patients anosognosics, from the term "anosognosia"—literally, an inability to recognize illness.) Ask them to raise their paralyzed arm, and they'll grin and claim it's tired, or say, "I've never been ambidextrous." Ask them to lift a tray full of cocktail glasses and—unlike normal hemiplegics, who grab the tray's middle with their one good arm—they'll grab the tray at one end, as if their other arm is there to help support it. When they (inevitably) flip the whole works over, they'll make up excuses. One anosognosic woman, when asked to clap, raised her one functioning arm and swished it in the air in front of her. (To which her doctor naughtily replied that she'd finally solved the old Zen koan: here was the sound of one hand clapping.) The really strange thing is how normal anosognosics seem otherwise. They can make jokes, reminisce about old times, and speak fluently. But their judgment has been warped, and on one topic especially—their disability—they're daft. A few stroke patients, after being stricken blind, deny even this deficit, to the detriment of their shins and the distress of anyone watching them cross the street.

We've learned a lot about the brain's hardware so far. Learned how neurons work and how they wire themselves together into circuits. Learned how webs of interlocking circuits create vision and movement and emotion. But it's time to cross over from the physical to the mental—and there's no better bridge for this than delusions. Doctors have dealt with delusions for millennia, of course, and have long known certain psychological facts: that many delusions disappear after a few weeks, and that some appear most commonly among certain personality types, such as perfectionists. But only in the past century have doctors seen enough cases, with enough eerily similar symptoms, to determine that many delusions have an organic basis in the brain. In fact, some delusions are so reproducible, and knock out such specific mental modules, that they've become a spectacular tool for probing one of the great mysteries of neuroscience: how cells and biochemicals give rise to the human mind, in all its oddness.

~

On June 3, 1918, a woman known only as Madame M. burst into a Paris police station, panting and near tears. She told the policeman on duty that she knew of at least 28,000 people, mostly children, being held hostage in Paris basements and catacombs. Some were being mummified alive, some were being flayed and experimented on by sadistic doctors, all were enduring unimaginable torture. When asked why no one had noticed this vast conspiracy, M. explained that every victim had been replaced by an aboveground "double"—a near-perfect replica who assumed the original person's identity. To verify her story, she demanded that two policemen accompany her immediately. They did—straight to an asylum.

M. had worked for years as a couture seamstress and designer, but for the psychologists who examined her, the salient part of her biography concerned her five children. Four had died as infants, including twin boys, a double blow that basically unsprung her mind. She began telling people that her little ones had been poisoned or abducted, and from there the fantasies only got wilder. M. spun tales of such complexity, in fact, that even she sometimes got lost in the labyrinth. She was supposedly descended from King Henri IV; but to erase her identity and cheat her out of her inheritance—including eighty billion francs and Rio de Janeiro—spies had dyed her flaxen hair chestnut, put drops in her eyes to change their size, and "stolen her breasts." How this meshed with the catacombs plot wasn't clear, and in general the tale held little interest for her doctor, Joseph Capgras: he'd seen plenty of lunatics invent grand genealogies for themselves. But one detail did strike Capgras as unusual, and significant—M.'s belief in doubles. She kept repeating that word, doubles, and insisted that even her last remaining family members, a daughter and her husband, had been murdered and replaced.

How had M. determined they were doubles? With the same skilled eye that had made her a couture seamstress. When relating stories to Capgras, M. would highlight the exact shade of ivory buttons on a garment, the exact type of satin lining in a coat, the exact kind of white feather adorning a hat. Similarly, with people, she might recall the precise hazel of someone's eyes and the precise length of men's mustaches, and she mapped people's scars and freckles with the same precision that ancient astronomers had charted the heavens. The problem is, people change: they get haircuts, nick their hands with knives, eat éclairs and gain a pound. And whenever the people in M.'s life changed, her brain would count them as a new person—a double—as if the "old" person had disappeared.

In fact, as the doubles themselves accumulated new wrinkles or went a little balder month by month, she confabulated doubles for them, and then doubles of the double-doubles. Eventually, she said, eighty doubles of her husband appeared. Her daughter was more promiscuous still, appearing in two thousand avatars between 1914 and 1918. There's no record of what happened to M., but in all likelihood she ended her sad life in an asylum.

After Capgras published his case report, other neuroscientists began to notice this doubling delusion in their patients, and Capgras syndrome is a well-recognized if rare ailment today. Most Capgras victims in decades past identified the imposters in their lives as actors or living waxworks; as new technologies emerged in the past century, the intruders became aliens, androids, and clones. Like M., some Capgras victims spun fabulous soap operas involving switched births and foiled inheritances. But just as often, victims complained about mundane things. One Capgras victim agonized to his priest that he'd committed bigamy, since he was now married to two women—his wife and her double. And not all the doubles were human. Some people sensed imposter cats and poodles. One person felt his hair had abandoned him, leaving an imposter wig.

As for the victims' relationships with the doubles, those varied.

Some accepted the interlopers. One sweet old woman began making three cups of tea each afternoon—for her, for her husband's double, and for her missing husband himself, just in case he returned that day. Other people found Capgras syndrome erotic. One Frenchwoman in the 1930s had complained for years about her awkward lover; luckily, his double proved a stud. Male victims liked that their wives' bodies seemed electrifyingly new every few weeks. (One cheeky doctor has even declared the syndrome the secret to connubial bliss, since each sexual encounter feels fresh.) Still, most Capgras victims fear the doubles and grow paranoid. And attempts to reason with them often backfire. Some loved ones have tried reminiscing with victims, by sharing details of their lives that only the two of them could know. But this proof of their genuineness can spook the victims, since the "imposter" obviously tortured this out of the missing person. A few victims have even killed doubles. A Missouri man decapitated his stepfather in the 1980s and then dug through his severed neck looking for the "robot's" batteries and microfilms.

To explain the origins of the syndrome, Capgras seized on one telltale fact: that victims can recognize their loved ones' faces, even while they deny it's "really" them. In other words, victims perceive people accurately but don't react appropriately to what they perceive—which implies that the root of the problem is emotional, since emotions help shape those reactions. Unfortunately, Capgras fell in with Freudians and decided to reinterpret his syndrome as a psychosexual neurosis (mostly a repressed desire for incest, natch). But doctors soon discovered that toxins, methamphetamines, bacteria, Alzheimer's disease, and blows to the brain could also induce Capgras syndrome, which weakened the Freudian theory. That accidents and disease could cause the syndrome suggested an organic basis, and neuroscientists eventually circled back to Capgras's prescient guess about emotions.

The full explanation of Capgras syndrome requires a quick trip back to face-blindness. The face-blind often cannot recognize even loved ones without using context clues or resorting to tricks. Nevertheless, many face-blind people do recognize faces on some level, no matter what they claim. Scientists have run experiments in which they gave a face-blind person—call him Chuck—a stack of Polaroids, some showing strangers, some showing loved ones. The scientists also placed electrodes on Chuck's skin to measure his emotional response to each picture. (Whenever someone experiences an emotion, his skin begins to sweat ever so slightly, even if he can't feel the moisture. Sweat contains dissolved salt ions, which increase the electrical conductivity of skin.*) When Chuck starts flipping through the Polaroids, he'll draw blanks on every face—don't know, don't know, still don't know. But his emotions do know. Whenever he perceives a loved one, the electrical flow on his skin rises by a measurable amount. His mind has no conscious access to the face's identity, but his subconscious bleats *dad, dad, dad.*

This cryptic emotional response implies that the human brain recognizes faces via two distinct circuits. Both rely on the brain's automatic analysis of lines and contours and other visual features. But while one circuit alerts us that this face is so-and-so, the other circuit bypasses this conscious route, plugging instead into our emotional centers and calling up the appropriate admiration or disgust. Fully recognizing a face, then, requires both conscious recognition and also what's called the "glow"—that ineffable connection we feel with someone else. Face-blind people get the glow, but because their visual recognition circuits are faulty, they have to rely on voice or some other clue to actually identify someone.

Now imagine the mirror image of face-blindness: imagine recognizing the face, but feeling no glow. That's Capgras syndrome. Give victims a stack of Polaroids, and their brains respond to loved ones and strangers with identical flatness. Even when they recognize Mom, their skin, and more to the point their hearts, feel no limbic tingle. That's not to say that Capgras victims are emotionally stunted. They can usually feel the full range of human emotions—for other stimuli.

Faces, though, cannot conjure up the proper feelings, and it's the chasm between what they once felt upon seeing a loved one and the deadness they now feel that inflicts the agony.

This dual-circuit theory of Capgras received a further boost from V. S. Ramachandran, the neuroscientist who developed the mirror-box therapy for phantom limbs (he has a thing for oddball neurology). Ramachandran was treating a thirty-year-old Brazilian man named Arthur who'd bashed his head on a windshield during a car accident. Arthur recovered his speech, memory, and reasoning skills, and never experienced any hallucinations or paranoia. But he confided to his doctors that someone had kidnapped and replaced his father. A rather intelligent man, Arthur knew on some level that this made no sense—why on earth would someone pretend to be his dad? Yet he couldn't shake the idea.

On a hunch, one day Ramachandran had Arthur's father walk down the hall and telephone his son, to isolate the effects of voice. To everyone's delight, the Capgras delusion evaporated. Father and son instantly reconnected—at least for the duration of the phone call. Once they met face-to-face again, Arthur's suspicions returned. Ramachandran traced this split to a simple anatomical fact. The brain routes both visual and auditory input into the limbic system for subconscious processing, but it uses different neural channels for each sense. Apparently, inside Arthur's brain, the vision-limbic circuit had suffered damage, while the hearing-limbic circuit had been spared. As a result, his father's voice retained its glow.

So why didn't Arthur feel the glow when he spoke to his father face-to-face? The short answer is that we dedicate so much of our brain to processing visual cues—half the cerebral cortex gets pulled in at various points—that vision overwhelms our other senses. So Arthur ignored his father's voice, however authentic, because in his eyes the man *looked* so uncannily sinister. Indeed, circumstances play an important role in Capgras delusions. Another way to think about Capgras is that it's a feeling of jamais vu, the inverse of déjà vu: instead of déjà vu's tantalizing familiarity in a strange context, Capgras victims sense a sinister strangeness in what should be a safe, familiar context.*

To me, Capgras is one of the most poignant neurological disorders out there. Other neurological diseases can sabotage people's ability to recognize loved ones, of course. But if dear Uncle Larry suffers from Alzheimer's and suddenly can't place you, most people accept that Larry isn't "there" on some level. Moreover, Larry doesn't single out any one person. A Capgras victim, meanwhile, seems fully there: his memory and speech and sense of humor remain intact, as do his emotions overall. He still adores the idea of you. But if you reach out to hug him, he'll reject you—reject you personally.

Beyond the emotional distress, Capgras can also plunge victims into existential quagmires. Consider those people who see doubles *of themselves,* especially lurking in the mirror. Oddly, these people understand how mirrors work; they realize that every other person on earth sees a genuine reflection there. They nevertheless insist that the mirror is lying in this special case: *that's a double of me.* As with Capgras generally, some people respond magnanimously to this intrusion. One man, although irked to find that his mirror-double always wanted to shave or brush his teeth at the same time, couldn't really hold a grudge against the imposter. Another noted that his double "wasn't a bad-looking fellow." More often, though, victims see a mirror-double as sinister: a stalker bent on replacing them. The families of some victims have to cover up mirrors and even reflective windowpanes with curtains, lest the victims catch an inadvertent glimpse and attack.

Above all, Capgras syndrome exposes a rift between reason and emotion inside the brain. We've already seen how reason and emotion can support each other. But they can also work at cross-purposes, and Capgras syndrome implies that, of the two, emotion is more primal and more powerful: why else would victims jettison all reason and invent doppelgangers and worldwide conspiracies just to explain a personal feeling of loss? Victims who don't acknowledge themselves in mirrors even argue for suspending the laws of physics. In some cases you wouldn't think your mind could survive such a rupture with reality. But it can: its defenses are ingenious, designed to silo your insanity to one topic and spare the mind in general.

~

While lounging around one afternoon in the spring of 1908, a middle-aged German woman felt an unseen hand grip her throat. She thrashed and gasped as it crushed her windpipe, and only after a great struggle did she manage to pry it loose with her right hand. At which point the offending hand—her own left hand—fell limply to her side. A few months prior, on New Year's Eve, she'd suffered a stroke, and ever since then her left hand had been lashing out like a rotten child—spilling her drinks, picking her nose, throwing off her bedcovers, all without her conscious consent. Now the hand had choked and bruised her. "There must be an evil spirit in it," she confessed to her doctor.

Two similar cases popped up in the United States during World War II. Both victims, one woman, one man, suffered from epilepsy and had had their corpus callosums surgically severed to head off seizures. (The corpus callosum, a bundle of neuron fibers, connects the left and right hemispheres.) The seizures did quiet down, but a distressing side effect emerged: one hand took on a life of its own. For weeks afterward the woman would open a drawer with the right hand and the left hand would snap it shut. Or she'd start buttoning up a blouse with her right and the left would follow along and unbutton it. The man found himself handing bread to his grocer with one hand, yanking it back with the other. Back at home, he'd drop a slice into the toaster and his other hand would fling it out—*Dr. Strangelove* meets *The Three Stooges.*

As more and more cases emerged, neurologists started calling this syndrome "capricious hand" and "anarchic hand," but most now refer to it as alien hand—the unwilled, uncontrolled movements of one's own hand. Alien hand can strike people after strokes, tumors, surgery, or Creutzfeldt-Jakob disease, and while cases usually disappear within a year, sometimes the hand anarchy persists for a decade.

Most cases of alien hand fall into one of two categories. The first involves "magnetic" clasping. A couch potato's hand snatches the remote and won't relinquish it. A pinochle player can't let go of a card she's dealing. A bingo player uses a nearby chair to pull himself to his feet and drags it all the way to the bathroom without realizing he's still holding on. That last incident seems unfathomable—how could he not know?—but more often than not, the victim remains oblivious to what his alien hand is doing until something bad happens. It's a spooky echo of the biblical command for one hand to keep secrets from the other.

The second type of alien hand pits righty against lefty in active opposition. One hand answers the phone, the other hangs up. One hand pulls your pants up, the other drops them to your ankles. And playing checkers? Fuggedaboutit—one hand repeatedly undoes the other's moves. In a variation, the offending hand might refuse orders: it won't dust part of the furniture or soap up half the torso in the shower. And in some victims, the two types of alien hand combine. One poor man, a seventy-three-year-old stroke victim with no history of sexual exhibitionism, would occasionally look down in public to find his fly open, his left hand going to town. And boy, once his hand clamped on, there was no letting go.

Many people refer to their alien hands as "imps" or "devils," and they often take harsh measures to control the mischief, up to and including beatings. Other victims pin their hands between a piece of furniture and the wall to trap them, or muzzle them in oven mitts. Often these measures fail, though—the hand pulls a Houdini—and some people live in constant terror of what it will do next. Alien hands have snatched boiling pots off the stove and grabbed at napkins on fire. They've swung axes, and suddenly yanked the steering wheel while driving. About the only known case of a benevolent alien hand involved a woman whose left hand would snap her cigarette case shut before she could get a smoke out.

Through autopsy work, neuroscientists have determined what sort of brain damage causes alien hand. First, victims probably suffer damage to sensory areas. Those areas provide feedback whenever we move our arms voluntarily, and without that feedback, people simply don't feel as if they've initiated a movement themselves. In other words, victims lose a "sense of agency"—a sense of being in control of their actions.

Magnetic grasping usually involves the dominant right hand and usually requires additional damage to the frontal lobes. The job description for the frontal lobes includes suppressing impulses from the parietal lobes, which are curious and capricious and, as the lobes most intimately involved with touch, want to explore everything tactilely. So when certain parts of the frontal lobe go kaput, the brain can no longer tamp down these parietal impulses, and the hand begins to flail and grab. (Neurologically, this flaring up of suppressed impulses resembles the "release" of the snout reflex in kuru victims.) And because the grasping impulse springs from the subconscious, the conscious brain can't always interrupt it and break the hand's grip.

Hand-to-hand combat—with one hand undoing the other's work (pants up/pants down)—usually arises after damage to the corpus callosum, damage that disrupts communication between the left and right hemispheres. The left brain moves the right side of the body, and vice versa. But proper movement involves more than just issuing motor commands; it also involves inhibitory signals. When your left brain tells your right hand to grab an apple, for instance, the left brain also issues a signal through the corpus callosum that tells your right brain (and thus, left hand) to cool it. The message is, "I'm on it. Take five." If the corpus callosum suffers damage, though, the inhibition signal never arrives. As a result the right hemisphere notices that something's going on and—lacking orders not to—lurches with the left hand to get in on the action. It's really an excess of enthusiasm. And because most people perform most tasks with their right hands, it's usually the left hand that jumps in late and causes this type of alien anarchy. Overall, if magnetic grasping usually involves the dominant half of the brain asserting its dominance even more, then left hand–right hand combat usually involves the weaker half rebelling and trying to win equal status for itself.

~

The presence of left/right conflict within the brain explains more than just alien hand. Hemispatial neglect usually arises after damage to the right hemisphere. Hence, Woodrow Wilson couldn't notice anyone to his left, and it's usually the left side of flowers and clocks that victims omit when doodling. The reason for this is cranial asymmetry. For whatever reason, the right hemisphere has superior spatial skills and does a better job mapping the world around us. So if the left brain falters, the right brain can compensate and monitor both sides of the visual field, thereby avoiding hemispatial neglect. The left brain, however, can't reciprocate: it can't make up for the loss of the right brain's spatial skills if the right brain falters. As a result, half the world disappears.

William O. Douglas's refusal to acknowledge his illness had a similar root cause. Douglas almost certainly suffered damage to areas in the right parietal lobe that monitor touch sensations such as pain, skin pressure, and limb position; without these touchy-feelies, it's hard to tell that your body parts aren't moving properly. Moreover, these right-brain areas also detect discrepancies. If you issue a command *(Lift the left arm),* and nothing happens because that arm is paralyzed, that's a discrepancy, and your right parietal lobe should send out an uh-oh alert. But if a stroke knocks out the uh-oh alert system, the brain will struggle to catch discrepancies, even blatant ones. It's like disabling a fire alarm. As a result, Douglas didn't recognize—in some sense couldn't recognize—that his whole left side couldn't move.

(In extreme cases, this lack of sensation and inability to detect discrepancies will lead a stroke victim to outright reject her paralyzed limbs. That is, she'll claim that she can't control her inert arm or leg because—despite its being attached *to her own body*—the limb actually belongs to someone else, a spouse or a mother-in-law, say. One victim, when shown her own wedding band on the very fingers she was disavowing, claimed the ring had been stolen. Another victim in a hospital complained that medical students kept slipping a cadaver arm under his sheets as a sick joke.)

Capgras syndrome also makes more sense with left/right discord in mind. The drastic conclusions of Capgras victims have always puzzled scientists. Losing the emotional connection to a loved one no doubt causes angst. But why confabulate imposters? Why doesn't logic intervene? The answer seems to be that full-blown Capgras actually requires two lesions: one to the face-emotion circuit, a second to (we're seeing a pattern here) the right hemisphere. According to this theory, the left and right hemispheres work together to help us understand the world. The right brain specializes in gathering sensory data and other simple facts. The left brain, meanwhile, prefers interpreting those data and spinning them into theories about how the world works. In a normal brain, there's a necessary give and take between these processes. For instance, if the left brain gets too fast and loose in forming a theory, the right brain can check it with a cold, hard fact and prevent a nutty idea from taking hold.

With Capgras, the sudden loss of emotional glow feels threatening and demands an explanation, which is the left brain's bailiwick. And if only the face-emotion circuit had suffered damage, the right brain would have supplied the relevant facts (this still looks like Dad, still talks like him) and guided the left brain to a sensible conclusion. When the right hemisphere suffers damage, though, that counsel disappears. So there's nothing to stop the left brain from twisting the facts to fit a preconceived notion. And given how cherished the belief that's being challenged is—whether you still feel love for Mom and Dad and your own children—it's no wonder that the brain prefers to spin tales of imposters and worldwide conspiracies rather than give it up. True, the conclusion seems to violate common sense, but common sense depends on intact brain circuits.

In light of brain discord many delusions seem, if not rational, at least comprehensible. They're simply the failings of a fragile brain. Sadly, though, explaining to a patient what causes his delusion rarely helps relieve it: given its nature, you can't talk someone out of a delusion so easily. (It's similar to how an optical illusion still fools us even when we know it's a trick. Our brains can't help it.) In fact, arguing with the victims of a delusion can backfire. Because if proved wrong, they'll often double down and blurt out something even wilder. *You tortured that memory out of my sister. I'm trying out for the Redskins.*

Some delusions run so deep that they fray the very fabric of the victim's universe. With so-called *Alice in Wonderland* syndrome—a side effect of migraines or seizures—space and time get warped in unsettling ways. Walls recede when approached, or the ground suddenly feels spongy beneath their feet. Worse, people feel themselves shrinking down to six inches in height or sprouting up to twelve feet tall. Or their heads feel swelled up, like cranial balloons. *Alice* victims* basically become the incarnations of fun-house mirrors, probably due to malfunctions in the parietal lobe areas responsible for body posture and position. Schizophrenics can experience severe delusions, too, like "delusional bicephaly"—what you might call Siamese twin disorder, the feeling of having an extra head. In 1978 an Australian schizophrenic killed his wife with his erratic driving. Two years later he suddenly found her gynecologist's noggin perched on his shoulder, whispering to him. Lord knows why, but the man took this as a sign that the gyno had diddled his wife, so he tried to guillotine the doctor's head with an axe. When this failed, he started shooting at the

Victims of *Alice in Wonderland* syndrome feel stretched or shrunk, much like Alice herself.

head with a gun and shot his own head by accident. (The subsequent brain damage from the bullet did "cure" him of this delusion.)

Perhaps the most absurd delusion—in the Sartre/Camus/existentialist sense of absurd—is Cotard syndrome, in which victims insist, absolutely swear, that they've died. Also known as walking dead syndrome, it usually strikes older women, and often emerges after an accident: they're convinced that their suicide attempts succeeded, or that they died in the car wrecks that sent them to the hospital. The seemingly blatant fact that they're sitting there, *telling you all this,* doesn't impinge; these are people who can hear Descartes's *cogito ergo sum* and say, Not so fast. Some even claim they can smell their own rotten flesh; a few have tried to cremate themselves. And in some cases, their delusions plumb the very depths of nihilism. As the first doctor to describe the syndrome, Jules Cotard, said: "You ask them their name? They don't have a name. Their age? They don't have an age. Where they were born? They were never born." Neurologists disagree about the explanation for Cotard, although most feel, as with Capgras syndrome, that two parts of the brain must be malfunctioning simultaneously. One theory interprets Cotard as Capgras turned inward: people feel no "glow" about themselves, and that deadness convinces them that they have in fact died, logic be damned.

All these delusions pry open the human mind and expose seemingly solid, seemingly unshakable aspects of our inner selves as really rather tenuous. Hemispatial neglect wipes out half the victim's world, and he never notices. Capgras victims lose the ability to feel close to people. *Alice* victims feel their bodies melt into instability. And alien hand syndrome upends our notions about free will, since victims seem to have lost free will for part of their bodies. But if the history of neuroscience proves anything, it's that any circuit for any mental attribute—up to and including our sense of being alive—can fail, if just the right spots suffer damage.

Like it or not, delusions can dupe even healthy brains. With nothing more than video cameras and mannequins, scientists can easily induce out-of-body experiences in volunteers. Or they can graft an extra arm onto someone's torso by simultaneously stroking both her real hand and a dummy hand attached to her. Some inspired setups can make people feel that they've changed genders or are shaking hands with themselves. *Hi, I'm Sam. Nice to meet you, Sam.*

Even more disarming are a series of experiments that began in San Francisco in the 1980s. A neuroscientist there named Benjamin Libet sat some college students (including his daughter) down in his lab and had them face a timer. He fit them with a helmetlike contraption that recorded the electrical activity in their brains, then told them to hold still. All the students had to do, for the entire experiment, was move one finger. Just whenever they felt like it: wait...wait...*tap.* Afterward they told Libet the precise moment on the timer when they'd decided to move. He then compared their answers to what the electrical scans said.

On every scan Libet could see a spike in motor activity not long before the finger moved. Pretty straightforward. The problem started when he looked at when the decision to move had taken place. Because in every case the conscious decision *lagged behind*—by a good third of a second—the unconscious spike in motor activity. Indeed, the spike was usually almost over before the decision got made. Because causes must precede effects, Libet concluded, reluctantly, that the unconscious brain must be orchestrating the whole sequence, and that the "decision" to move was nothing but a post hoc rationalization—an ego-saving declaration by the conscious brain. *Uh, I meant to do that.* This experiment has been replicated many times—it's robust. And in many cases scientists can predict when someone will move before even she knows that.

Equally unnerving is another set of experiments, which involved stimulating the exposed brains of surgical patients with electricity. When scientists sparked certain motor areas, people's arms and legs flailed. But unless the person actually saw himself move, he denied he'd done so, since he'd felt no inner urge to. Conversely, sparking other parts of the brain can induce just the urge, even while the arms and legs lie limp. Stronger currents can even induce a false sense of having moved, but again without any actual movement taking place. (One woman said, in all seriousness, "I moved my mouth there, I talked. What did I say?") In sum, your actions, your desires to act, and your conviction of having acted can all be decoupled and manipulated. None of those three things necessarily follows from the others; they're more casually linked than causally linked.

If you're biting your nails and wondering where free will fits into all this, you're not alone. These experiments leave little wiggle room, and to many scientists they in fact obliterate free will. In this thinking, the mind's conscious, decision-making "will" is actually a by-product of whatever the unconscious brain has already decided to do. Free will is a retrospective illusion, however convincing, and we feel "urges" to do only what we're going to do anyway. Pride alone makes us insist otherwise. And if that's true,* victims of alien hand and other syndromes may have simply lost the illusion of free will for part of their bodies. In some sense, they might be closer to the reality of how the brain works than the rest of us. Makes you wonder who's really deluded.

PART V

CONSCIOUSNESS

CHAPTER TEN

Honest Lying

Almost every structure we've examined so far contributes to forming and storing memories. Memory is therefore a wonderful way to see how different parts of the brain work together on a large scale.

Soldiers buried more than men in the graves of Southeast Asia. While conquering Singapore in February 1942, Japanese soldiers captured 100,000 mostly British POWs, more than they knew what to do with. The military worked thousands of them to death on the brutal Burma–Siam "Death Railway," a project that required hacking through 250 miles of mountainous jungles and constructing bridges over rivers like the Kwai. Most of the remaining captives, including many doctors, were crowded into the notorious Japanese prison camps. In fact, two British doctors incarcerated in the Changi camp, Bernard Lennox and Hugh Edward de Wardener, realized that their captors were essentially running a gruesome experiment: taking healthy men, depriving them of one nutrient, and watching their brains deteriorate.

No matter his background, every doctor in the camps worked as a surgeon, dentist, psychiatrist, and coroner, and they suffered from the same ailments—dysentery, malaria, diphtheria—that ravaged the troops. They pared down bamboo shards for needles, unstitched parachutes for silk sutures, and drained human stomachs for acids. Monsoons tore through their "clinics"—often just tents draped over poles—and some doctors faced beatings and threats of being boiled in oil if they didn't cure enough soldiers to meet work quotas. Guards made things worse by restricting sick men to half rations, to "motivate" them to recover. But even among the healthy, the food—mostly plain rice—was never adequate, and led to beriberi disease.

For as long as people have eaten rice in Asia, doctors there have reported outbreaks of beriberi. Symptoms included heart trouble, anorexia, twitching eyes, and legs so swollen that the skin sometimes burst. Victims also walked with a shuffling, staggering gait that reminded locals of *beri,* sheep. When Europeans colonized Southeast Asia in the 1600s, their doctors began seeing cases as well; one early report came from Dr. Nicolaes Tulp, the Dutchman later immortalized in Rembrandt's *The Anatomy Lesson.* But the number of cases exploded after the introduction into Asia, in the later 1800s, of steam-powered rice mills. The mills removed the outer husks from rice grains, producing so-called white rice. People back then called it polished rice, and cheap polished rice became a dietary staple—or, often, *the* diet—of peasants, soldiers, and prisoners. During the Russo-Japanese War alone, 200,000 Japanese troops fell victim to beriberi.

Scientists eventually began to suspect that beriberi was a nutritional deficiency—probably a lack of vitamin B_1 (a.k.a. thiamine). In shucking off the nutritious rice husks, the mills stripped out almost all the B_1, and many people didn't get enough thiamine from eating vegetables, beans, or meat. Our bodies use B_1 to harvest energy from glucose, the end result of digesting carbohydrates. Brain cells especially rely on glucose for energy, since other sugars cannot cross the blood-brain barrier. The brain also needs thiamine to make myelin sheaths and to build certain neurotransmitters.

The first cases of beriberi appeared two weeks after the Changi camp opened, among a few alcoholics cut off cold turkey. Many more cases appeared after another month. Doctors tended to the ailing as best they could and sometimes kept their spirits up by lying about the

A hospital in a Japanese POW camp in Singapore.

progress of Allied armies. When all else failed, some doctors ordered men to live or face court-martial (a threat reminiscent of those old medieval laws that made suicide illegal). Nevertheless, by June 1942 there were a thousand beriberi cases in Changi alone. Helpless to stop the epidemic, de Wardener and Lennox started doing autopsies in secret and collecting tissues from the brains of beriberi victims, to study the pathology of the disease.

Although considered contraband, these tissues and autopsy records were mostly safe inside Changi. But in 1943 Lennox and de Wardener were herded off to different camps near the Death Railway in Siam and had to split their medical stash. Wary of confiscation, Lennox arranged to smuggle the brain tissues out of his camp, only to have them perish in a train wreck. De Wardener guarded the all-important paper records, a four-inch sheaf. But as the war turned sour for Japan in early 1945, de Wardener realized that Japanese leaders wouldn't look kindly on hard evidence of starving POWs. So when he received another transfer order—and saw guards frisking his fellow-transferees and searching their belongings—he made a hasty decision. He had a metallurgist friend seal his papers inside a four-gallon petrol tin. He then wrapped the tin in a cape and buried the bundle three feet deep in a fresh grave, leaving only the dead soldier as a sentinel. To remember which grave it was—there were so many—he and some friends took compass bearings on a few enormous trees nearby. As he departed camp, de Wardener could only pray that the heat, rot, and miasma of Siam wouldn't eat through the bundle before he returned. If he returned.

The records were precious because they resolved a half-century-long dispute about the brain, B_1, and memory. In 1887 a Russian neuroscientist named Sergei Korsakoff described a peculiar ailment among alcoholics. Symptoms included emaciation, staggering, a lack of the patellar kick reflex, and urine "as red as the strongest tea." But the outstanding symptom was memory loss. Korsakoff's patients could play chess, banter, make wisecracks, and reason properly—but couldn't remember the previous day, even the previous hour. During conversations they repeated the same anecdotes over and over, verbatim. And if Korsakoff left the room for a spell, they repeated the same anecdotes over and over, verbatim, when he returned. Other brain diseases cause memory loss, of course, but Korsakoff noticed something distinctive about these cases. If asked a question they can't answer, most people with memory loss admit they don't know. Korsakoff's patients never did—they always lied instead.

Today, Korsakoff's syndrome—the tendency to lie compulsively due to brain damage—is a well-recognized ailment. And truth be told, it can be quite entertaining, in a gallows-humor way. When asked why Marie Curie was famous, one Korsakoff victim declared, "Because of her hairstyle." Another claimed to know Charlemagne's favorite meal ("maize porridge") and what color horse King Arthur rode ("black"). Victims lie especially often about their personal lives. One man claimed to remember, thirty years later, what he wore the first day of summer in 1979. Another told his doctor, in consecutive sentences, that he'd been married for four months and that he'd sired four children with his wife. After a quick calculation, he marveled at his sexual prowess: "Not bad."

Beyond the occasional Münchhausenian whopper, most Korsakoff victims tell plausible, even mundane lies: unless you knew their life histories, you'd never peg them as bullshit artists. Unlike most of us, they don't lie to make themselves look good, or to get an edge, or to conceal something. And unlike people suffering from delusions, they don't defend themselves ferociously if called out; many just shrug. But no matter how many times someone catches them, they keep lying. This fibbing for no obvious or underhanded reason is known as confabulation.

Korsakoff focused on the psychology of confabulation, but other scientists extended his work in the early 1900s and started linking these psychological symptoms to specific brain damage. In particular, they discovered tiny hemorrhages in the brains of victims, as well as patches of dead neurons. Pathologists also linked Korsakoff's syndrome to another, related disease called Wernicke's syndrome. In fact, because Wernicke's syndrome often turns into Korsakoff's, the two were eventually yoked together as Wernicke-Korsakoff syndrome.

The underlying cause of Wernicke-Korsakoff syndrome took longer to suss out, but by the later 1930s a few scientists had linked it to a lack of B_1. As doctors now know, alcohol prevents the intestines from absorbing the thiamine in food. This shortage then causes changes inside the brain, especially to glial cells. Among other jobs, glial cells sponge up excess neurotransmitters from the synapses between neurons. And without thiamine, the glia cannot sop up glutamate, which stimulates neurons. As a result of this excess, neurons get overstimulated and eventually exhaust themselves, dying of excitotoxicity.

Because they seemed to share a common root—B_1 deficiency—beriberi and Wernicke-Korsakoff syndrome should have caused similar symptoms and similar destruction inside the brain. But through the 1940s no one had any hard evidence to link them. This was partly because Wernicke-Korsakoff remained rare and associated primarily with alcoholics, and partly because doctors who studied beriberi focused on nerve and heart damage, not brain damage. The net result was confusion: was this one disease or two? More important, it highlighted a growing concern over efforts to link physiology and psychology: many doctors frankly doubted that the lack of a simple vitamin—a molecular problem—could leap up so many levels of scale and cause complex mental troubles like confabulation.

Changi proved it could. Among the thousand-plus beriberi victims there, several dozen also came down with symptoms of Wernicke-Korsakoff syndrome, including confabulation. As an example, de Wardener asked one far-gone man, just to test his mental state, "Do you remember when we met in Brighton? I was riding a white horse and you a black horse, and we rode on the beach." This was bunk, but the man answered that of course he remembered, and filled in the details. Such imaginings often became the patients' reality, sadly, and a few men died in this state—their last "memories" nothing but vapors and fabrications. Medically, the fact that beriberi always preceded Wernicke-Korsakoff, and that those with the worst beriberi got the worst Wernicke-Korsakoff, implied a common cause. Autopsies then cemented the link: even without a microscope, Lennox, a trained pathologist, could see the characteristic hemorrhages and patches of dead neurons in the brains of victims. Beriberi and Wernicke-Korsakoff seemed to be two stages, chronic and acute, of the same underlying disease.

As further evidence, treating victims with pure thiamine (some doctors had tiny stashes) usually relieved the symptoms of both Wernicke-Korsakoff and beriberi, sometimes within hours: de Wardener remembers a few men roaring to life and consuming whole mountains of rice to combat their sudden hunger. (Mental symptoms such as confabulation might take several weeks to dissipate.) For less acute cases doctors might add Marmite to meals (however unappetizing, this yeast-based extract is lousy with B_1) or ferment rice and potatoes to cultivate wild yeast, also chock-full of B_1. Some doctors sent men to gather thiamine-rich hibiscus leaves as well. The smarter doctors lied to the men and claimed that hibiscus would pump up their libidos for when they got back home to their gals. After that, troops no doubt couldn't consume enough hibiscus.

In tandem, the fact that consuming too little thiamine provoked Wernicke-Korsakoff, and that restoring thiamine to the diet relieved it, convinced Lennox and de Wardener that the lack of a simple nutrient could indeed destroy something as profound as our memories, even our sense of truth. But the duo still had to make their case to the medical world—which meant not only surviving the camps but preserving their autopsy files. This wasn't easy in a war zone, and as de Wardener discovered, about the only way to conceal such things was to bury them, and pray to God they survived.

After V-J Day, de Wardener received mysterious orders to report to Bangkok. Although anxious to start searching for his files, he remembers enjoying the journey: "I took a victorious ride across Siam in a Jeep, with all the Nips bowing... which was very satisfying." To his surprise, he found his records waiting for him at Bangkok HQ. Apparently a friend had returned to Changi with a shovel not long before, scrabbled through the dirt above the dead sentinel's body, and liberated the bundle. It was a close thing: the cape had rotted away and the solder sealing up the tin had disintegrated. But the papers had survived, perhaps by a matter of days. Lennox and de Wardener finally published this, well, groundbreaking work in 1947.*

~

Since World War II, neuroscientists have continued to mine confabulation for insight into how memory works, and it has proved a rich vein indeed. For example, confabulations reveal that each memory seems to have a distinct time stamp, like a computer file. And just like computer files, that time stamp can be corrupted. Most confabulators tell plausible lies; in fact, many of their false "memories" did happen to them at some point. But confabulators often mistake *when* the memory happened: the scenes in their lives have been shuffled wrong. So while they claim they ate truffled duck last night, in truth they did that thirty years ago while honeymooning in Paris. In some sense, then, confabulation is a breakdown in the ability to tell a coherent story about our lives.

The fact that virtually all confabulators have frontal lobe damage also tells us something. The frontal lobes help coordinate multistep processes, and despite how effortless memory seems, remembering something specific (say, the worst Christmas present you ever got) is complicated. The brain has a fraction of a second to search for the memory, retrieve it, replay it, and summon up the proper sensations and emotions—and that's assuming you recorded the memory accurately in the first place. If the frontal lobes suffer damage, any one of those steps can go awry. Perhaps confabulators simply retrieve the wrong memory each time they "recall" something, and don't recognize their error.

Some scientists trace confabulations to shame and a need to cover up deficiencies. Confabulators don't generally blurt things out unprovoked; you have to ask questions to elicit the lie. And according to this theory, admitting they don't know something upsets and embarrasses people, so they pretend. For example, most doctors ask at intake how many children someone has. Having to admit "I don't know" could be catastrophic to a person's well-being, since what kind of monster doesn't remember his own children? In short, confabulations could be a defense mechanism, a way for people to hide their brain damage, even from themselves.

As another defense mechanism, some confabulators invent fictional characters and foist their personal failings onto them. One alcoholic confabulator raved to his doctor about imps who kept breaking into his apartment, even after he changed the locks, and stealing things like his remote control. He eventually heaved the imps outdoors on a brutal January night. But, feeling guilty, he braved the weather and draped clothes over them later that night, then called an ambulance. In reality medical workers had discovered *him* outside that winter, stone drunk and mostly naked. In telling the story he was basically confabulating an allegory on the fly. That's a remarkable deed for someone with brain damage, and the ruse allowed him to ponder his own flaws more objectively, without implicating himself.

As that last case shows, it's not always clear whether confabulators understand that they're lying. Most seem blithely unaware, and many neuroscientists insist that Korsakoff patients don't realize what's happening. But is that possible? Covering up a memory gap, even subconsciously, implies that they know on some level that the gap exists. Which means they know and don't know at the same time. It's a doozy of a conundrum, and it raises all sorts of stoner questions about whether you can truly deceive yourself, and more broadly about the nature of truth and falsehood. Consider asking a confabulator what she ate for breakfast. If she hasn't the foggiest, she might blurt, "Leftover pizza." But of course it's possible she did have cold pizza for breakfast, in which case she would be telling the truth — even though her brain tried, consciously or not, to put one over on you. What on earth would you call that? Neither *lying* nor *telling the truth* quite encompasses it. It's slipperier, and some neuroscientists have taken to calling it "honest lying."

Philosophical conundrums aside, work on confabulation helped make memory a proper object of neuroscientific study last century, since scientists could finally link memory to the brain and its biology. That said, the biggest breakthrough in memory research in the past hundred years didn't spring from the minds of confabulators. Indeed, most memory work until the 1950s relied on a flawed assumption — that all parts of the brain contribute equally to forming and storing memories. It's an idea that took something drastic, a botched operation by a lobotomist, to overturn.

~

In the early 1930s a bicyclist in Connecticut struck a small boy, who tumbled and cracked his skull. No one knows whether the accident alone caused his epilepsy — three cousins had it, so he might have been predisposed — but the blow probably precipitated it, and at age ten he started having seizures. Each lasted around forty seconds, during which time his mouth flopped open, his eyes slipped shut, and his arms and legs crossed and uncrossed as if curled by an invisible puppeteer. He suffered his first grand mal on, of all days, his fifteenth birthday, while riding in the car with his parents. More followed, in class and at home and while shopping — up to ten seizures a day, with at least one major episode per week. So at an age when most people are struggling to find an identity, he was saddled with one he didn't want: the kid who shook, who bit his tongue, who slumped over and blacked out and pissed himself. The mockery got so bad he dropped out of high school, and he earned his diploma only at twenty-one, from a different school. He ended up living at home and working in a motor shop.

Finally the desperate young man — soon immortalized as H.M. — decided to try surgery. When younger, H.M. had dreamed of practicing neurosurgery himself and studying how the brain works. But while H.M. did end up contributing, profoundly, to neuroscience, his affliction ensured that he would never grasp his own importance.

H.M. started seeing Dr. William Scoville around 1943. A noted daredevil — before a medical conference in Spain once, he'd stripped off his jacket and mixed it up with the *toros* in the bullring — Scoville liked risky surgeries, too, and had jumped onto the American lobotomy bandwagon* early. But he disliked the drastic changes in his patients' personalities, so he began experimenting with "fractional" lobotomies, which destroyed less tissue. Over the years he basically worked his way around the brain, carving out this piece or that and checking the results, until he finally reached the hippocampus.

Because it was part of the limbic system, scientists at the time believed that the hippocampus helped process emotions, but its exact function remained unknown. Rabies often destroyed it, and James Papez had singled it out for attention. (A poetaster, Papez even penned a ditty to his wife that read: "It's Pearl, my girl on Broad Street / that I

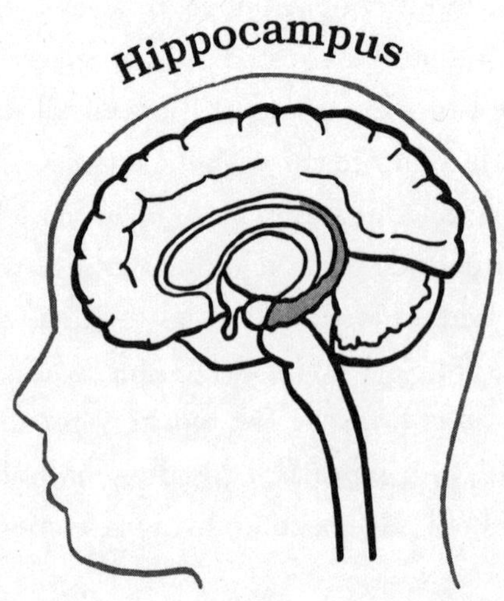

miss . . . My hippocampus tells me this.") Scoville was less enamored: he'd seen the mental turmoil that hippocampus damage could cause. So in the early 1950s he started removing the hippocampi (you have one in each hemisphere) from a few psychotics. Although it was hard to be sure in people with such disturbed minds, they seemed to suffer no ill effects, and two women in particular showed a marked reduction in seizures. Unfortunately Scoville neglected to do careful follow-up tests until November 1953—after he'd convinced H.M. to try the surgery.

H.M.'s operation took place in Hartford, Connecticut, on September 1, 1953. Scoville peeled back his patient's scalp, then used a hand crank and one-dollar drill saw from a local hardware store to remove a bottle cap's worth of bone from above each eye. As cerebrospinal fluid drained away, the brain settled down in its cavity, giving Scoville more room to work. With what looked like an elongated shoehorn, he nudged aside H.M.'s frontal and temporal lobes and peered inside.

The hippocampus sits at ear level and has the rough shape and diameter of a curled thumb. Hoping to remove as little tissue as possible, Scoville first sparked each hippocampus with wires to find the origin of H.M.'s seizures. No luck, so he grabbed a long metal tube and began cutting and sucking out tissue gram by gram; he eventually removed three inches' worth of hippocampus on each side. (Two nubs of hippocampal tissue remained behind, but because Scoville also removed the connections between those nubs and other parts of the brain, the nubs were useless, like unplugged computers.) For good measure, Scoville removed H.M.'s amygdalae and other nearby structures as well. Given how deeply all these structures are embedded in the brain, only a neurosurgeon could have destroyed them with such precision.

Post-op, H.M. remained drowsy for a few days, but he could recognize his family and carry on a seemingly normal conversation. And by many measures, the operation succeeded. His personality never changed; the seizures all but disappeared (two attacks per year at most); and when the fog of epilepsy lifted, his IQ jumped from 104 to 117. Just one problem: his memory was shot. Aside from a few small islands of recollection—like the fact that Dr. Scoville had operated on him—an entire decade's worth of memories from before the surgery had vanished. Equally terrible, he couldn't form new memories. Names escaped him now, as did the day of the week. He repeated the same comments over and over, verbatim, and while he might remember directions to the bathroom long enough to get there, he always had to ask again later. He'd even consume multiple lunches or breakfasts if no one stopped him, as if his appetite had no memory, either. His mind had become a sieve.

In light of modern knowledge, H.M.'s deficit makes sense. Memory formation involves several steps. First, neurons in the cortex jot down what our sensory neurons see and feel and hear. This ability to record first impressions still worked in H.M. But like messages scrawled on the beach, these impressions erode quickly. It's the next step, involving neurons in the hippocampus, that makes memories last. These neurons produce special proteins that encourage axon bulbs to swell in size. As a result, the axons can stream more neurotransmitter bubbles toward their neighbors. This in turn strengthens the synapse connections between those neurons before the memory decays. Over months and years—provided the first impression was strong enough, or we think about the event from time to time—the hippocampus then transfers the memory to the cortex for permanent storage. In short, the hippocampus orchestrates both the recording and the storage of memories, and without it, this "memory consolidation" cannot occur.

Scoville couldn't have known all this, but he'd clearly sabotaged H.M.'s memory, and he didn't know what to do. So a few months later, when he saw that Wilder Penfield was about to publish a report on hippocampus damage, Scoville called the renowned surgeon and confessed.

Penfield had recently operated on two patients with hippocampal epilepsy. To be safe, he'd removed the structure on just one side, but unbeknownst to him, the seizures had already destroyed the other hippocampus in each person. So removing the one left both patients without a working hippocampus, and they developed the purest amnesia Penfield had ever seen. Although he was still puzzling through the cases, a graduate student was going to present them at a scientific meeting in Chicago in 1954.

When Scoville called, Penfield reportedly flipped out, berating him for his recklessness. After calming down, though, the scientist in Penfield realized (much as the beriberi doctors had) that Scoville had actually performed an invaluable experiment: here was a chance to determine what the hippocampus did. As part of its mission Penfield's clinic in Montreal tracked the psychological changes that patients experienced after psychosurgery. So Penfield dispatched a Ph.D. student from the Neuro, Brenda Milner, down to Connecticut to investigate the hippocampusless H.M.

After his memory vanished, H.M. lost his job and had no choice but to keep living with his parents. He spoke in a monotone now and had no interest in sex, but otherwise seemed normal. To the neighbors, it probably just looked like he was loafing his life away. He took a part-time job packing rubber balloons into plastic bags, and did odd chores around the house. (Although his parents had to remind him where they kept the lawn mower every single time, he could actually mow just fine, since he could see what grass he hadn't cut.) His temper did flare up occasionally: his mother tended to nag, and he cuffed her a few times and kicked her shins. Another time, when an uncle removed a few choice rifles from the family's gun collection, he flew into a rage. (Despite his amnesia he retained a lifelong love of guns, and always remembered to renew his NRA membership.) But he whiled away most days peacefully, either doing crossword puzzles—working through the clues methodically, in order—or flopping in front of the television and watching either Sunday Mass or the old movies that, to him, would never become classics. It was like early retirement, except for the days Milner arrived to test him.

Milner would take the night train down from Montreal to Hartford, arriving at 3 a.m. and spending the next few days with H.M. Her battery of tests confirmed Scoville's basic observations pretty quickly: H.M. had little memory of the past and no ability to form new memories going forward. This was already a big advance—proof that some parts of the brain, namely the hippocampus, contribute more to forming and storing memories than other parts. And what Milner discovered next redefined what "memory" even meant.

Rather than keep asking him questions he couldn't answer, she started testing H.M.'s motor skills. Most important, she gave him a piece of paper with two five-pointed stars on it, one nested inside the other: ☆. The outer star was about six inches wide, and there was a half-inch or so gap between them. The test required H.M. to trace a third star between the two with a pencil. The catch was, he couldn't see the stars directly: Milner had shielded the diagram, and he had to look at them in a mirror instead. Left was right, right was left, and every natural instinct about where to move his pencil was wrong. Anyone taking this mirror test for the first time makes a mess—the pencil line looks like an EKG—and H.M. proved no exception. Somehow, though, H.M. got better. He didn't remember any of the thirty training sessions Milner ran him through. But his unconscious motor centers did remember, and after three days he could trace the star in the mirror fluently. He even commented near the end, "This is funny...I would have thought it would be rather difficult, but it seems I've done pretty well."

Milner remembers the star test as a eureka. Before this, neuroscientists thought of memory as monolithic: the brain stored memories all over, and all memory was essentially the same. But Milner had now teased apart two distinct types of memory. There's declarative memory, which allows people to remember names, dates, facts; this is what most of us mean by "memory." But there's also procedural memory—unconscious memories of how to pedal a bicycle or sign your name. Tracing the stars proved that H.M., despite his amnesia, could form new procedural memories. Procedural memories must therefore rely on distinct structures within the brain.

This distinction between procedural and declarative memories (sometimes called "knowing how" versus "knowing that") now undergirds all memory research. It also sheds light on basic mental development. Infants develop procedural memory early, which explains why they can walk and talk fairly quickly. Declarative memory develops later, and its initial weakness prevents us from remembering much from early childhood.

Another distinct type of memory emerged from Milner's tests as well. One day Milner asked H.M. to remember a random number, 584, for as long as possible. She then left him alone for fifteen minutes while she had a cup of coffee. Contrary to her expectation, he still knew the number when she returned. How? He'd been repeating it under his breath, over and over. Similarly, H.M. could remember the words "nail" and "salad" for several minutes by imagining a nail piercing some salad greens and reminding himself over and over not to eat the impaled leaves. Any distraction during those minutes would have ejected the words clean out of H.M.'s mind, and five minutes after the test ended, even the memory of having to remember something had vanished. Nevertheless, as long as H.M. concentrated and kept refreshing his memory, he could hold on. This was the first clue that short-term memory exists; moreover, it showed that short-term memory (which H.M. had) and long-term memory (which he lacked) must utilize different brain structures.

After Milner's discoveries, H.M. became a scientific celebrity, and other neuroscientists began clamoring to explore his unique mind. He did not disappoint. In April 1958, five years after the operation, H.M. and his parents moved into a small Hartford bungalow. In 1966 a few American neuroscientists asked him to draw the home's floor plan from memory. He succeeded. He didn't know the bungalow's address, but walking through its six rooms over and over had tattooed the layout into his brain. This proved that our spatial memory systems, while normally reliant on the hippocampus, can circumvent it if need be (probably via the parahippocampus, a nearby navigation center).

Scientists also discovered that time worked differently for H.M. Up to about twenty seconds, he reckoned time as accurately as any normal person. After that, things veered wildly. Five minutes lasted, subjectively, just forty seconds for him; one hour lasted three minutes; one day fifteen minutes. This implies that the brain uses two different timekeepers—one for the short term and one for everything beyond twenty seconds, with only the latter suffering damage in H.M. Once again, H.M. allowed scientists to break a complex mental function down into different components and to link those components to structures in the brain. Eventually more than one hundred neuroscientists examined H.M., making his probably the most studied mind in history.

All the while H.M. got older, at least physically. Mentally, he remained stuck in the 1940s. He remembered not a single birthday or funeral after that time; the Cold War and sexual revolution never registered; new words such as *granola* and *Jacuzzi* remained forever undefined. Worse, a vague sense of uneasiness often bubbled up inside him, and he could never quite shake it. The feeling, Milner reported, was "like that fraction of a second in the morning, when you are in a strange hotel room, before it all falls in[to] place." Only for H.M. it never did.

In 1980, after H.M.'s father died and his mother got too sick to care for him, he moved into a nursing home. He walked a little gimpily by that point: years of taking heavy-duty epilepsy drugs had withered his cerebellum, and his wide, shuffling gait resembled that of kuru victims. He also got pretty portly after too many forgotten second helpings of cake and pudding. But overall he was a fairly normal patient and lived a (mostly) placid life. He loafed through the nontesting days reading poems or gun magazines, watching trains rumble by, and petting the dogs, cats, and rabbits the facility owned. He learned how to use a walker, thanks to his intact motor memories, and he even attended his thirty-fifth high school reunion in 1982. (Although he recognized no one there, other attendees reported the same problem.) When he dreamed at night, he often dreamed of hills—not of struggling up them, but cresting them and being at the top.

Still, the old, volatile H.M. did flare up now and again. He sometimes refused to take his meds—at which point his nurses scolded him, warning him that Dr. Scoville would get angry if he disobeyed. (That Scoville had died in a car crash didn't matter. H.M. always fell for it.) He got into fights with other residents as well. One harpy at the nursing home would erase his bingo card midgame and taunt him. H.M. sometimes responded by running to his room and either banging his head on the wall or grabbing his bed and shaking it like a gorilla would its cage. One fit got so violent that his nurses called the police. These were moments of pure animal frustration—and yet in some ways they seem like his most human moments. For a few seconds a real person broke through the dull, bovine exterior. He was reacting the way we'd all want to if dealt his fate: he raged.

As soon as a nurse distracted H.M., he forgot his torment, of course. And aside from those flare-ups he lived a quiet life, albeit in declining health. He finally died in 2008, aged eighty-two, of respiratory failure—at which point scientists revealed him to the world as Henry Gustav Molaison.

The world of neuroscience mourned Molaison: his death led to numerous tributes about his patience and kindness, as well as scores of puns about his being unforgettable. And his brain is still providing insight today. Before his death, his nursing home had started stockpiling ice packs in preparation; when he passed, employees ringed his skull with them to keep his brain cool. Doctors soon arrived to claim the body, and that night they scanned his brain in situ and then liberated it. After two months hardening in formalin, it was flown cross-country in a cooler (which got the window seat) to a brain institute in

The brain of H.M., the unforgettable amnesiac, being sliced in preparation for future study. (Courtesy Jacopo Annese, the Brain Observatory, San Diego)

San Diego. Scientists there soaked it in sugar solutions to draw out excess water, then froze it to solidify it. Finally, they used the medical equivalent of a deli slicer to shave Molaison's brain into 2,401 slices, each of which they mounted on a glass plate and photographed at 20x magnification, to form a digital, zoomable map down to the level of individual neurons. The slicing process was broadcast live online, and 400,000 people tuned in to say goodbye to H.M.

~

Although H.M. dominated the scientific literature and popular imagination, plenty of other amnesiacs have contributed to our understanding of memory. Take K.C., an amnesiac in suburban Toronto. During a wild and extended adolescence, K.C. jammed in rock bands, partied at Mardi Gras, played cards till all hours, and got into fights in bars; he was also knocked unconscious twice, once in a dune buggy accident, once when a bale of hay konked him. Finally, in October 1981, at age thirty, he skidded off an exit ramp on his motorcycle. He spent a month in intensive care and lost, among other structures, both hippocampi.

After the accident a neuroscientist named Endel Tulving determined that K.C. could remember certain things just fine. But everything he remembered fell within one restricted category: it was all stuff you could look up in reference books, like the difference between stalactites and stalagmites or between spares and strikes in bowling. Tulving called these bare facts "semantic memories," memories devoid of all context and emotion.

At the same time K.C. had zero "episodic memory"—no memories of things he'd personally done or felt or seen. For instance, in 1979 K.C. surprised his family the night before his brother's wedding by getting a perm. To this day he knows his brother got married and can recognize family members in the wedding album (the facts), but he doesn't remember being at the wedding and has no idea how his family reacted to his curly hair (the personal experiences). The little that K.C. did retain about his preaccident life sounds like something he looked up in a particularly dry biography of himself. Even pivotal moments have been reduced to bulleted points in an index. He knows his family had to abandon his childhood home because a train derailed and spilled toxic chemicals nearby; he knows a beloved brother died two years before his own accident. But these events have no emotional import anymore. They're just stuff that happened.

These details, along with scans of K.C.'s brain, provided strong evidence that our episodic and semantic memories rely on different brain circuits. The hippocampus helps record both types of memories initially, and it helps retain them for the medium term. The hippocampus probably also helps us access old *personal* memories in long-term storage. But to access old semantic memories, the brain seems to use the parahippocampus, an extension of the hippocampus on the brain's southernmost surface. K.C., whose parahippocampi survived, could therefore remember to sink the eight ball last in pool (semantic knowledge), even though every last memory of playing pool with his buddies had disappeared (personal knowledge).*

What's more, while a healthy hippocampus will usually take responsibility for recording new semantic memories, the parahippocampus can — albeit excruciatingly slowly — absorb new facts if it has to. For instance, after years of shelving books as a volunteer at a local library, K.C.'s parahippocampus learned the Dewey decimal system, even though he had no idea why he knew it. Similarly, H.M.'s healthy parahippocampus picked up a few choice facts after his 1953 surgery. After seeing the crossword clue a thousand times he dimly recalled that "Salk vaccine target" equaled P-O-L-I-O. And through incessant references, he retained a sliver of information about the 1969 moon landing and 1963 Kennedy assassination. Contra the cliché, he couldn't recall where he was when he learned those things — that's episodic memory. And his knowledge of the events remained weak and fragmentary, since the parahippocampus cannot learn very well. He nevertheless absorbed that they'd happened.

Along these same lines, K.C. helped neuroscience come to grips with another important distinction in memory research, between recollection and familiarity. Colloquially, recollection means *I specifically remember this,* while familiarity means *this sounds familiar, even if the details are fuzzy.* And sure enough, the brain makes the same distinction. In one test K.C.'s doctors compiled a list of words (El Niño, posse) that entered the common parlance after his accident in 1981. They then sprinkled those words into a list of pseudo-words — strings of letters that looked like plausible words but that meant nothing. Time and again K.C. picked out the real word, and did so with confidence. But when asked to define the word, he shrugged. From a list of common names he picked out the people who'd become famous after 1981 (e.g., Bill Clinton). But he had no inkling what Clinton had done. In other words, K.C. found these terms familiar, even though specific recollection eluded him. This indicates that recollection once again requires the hippocampus, while a feeling of familiarity requires only certain patches of cortex.

A final type of memory that amnesiacs have helped illuminate is emotional memory—which makes sense, given that the hippocampus belongs to the limbic system. Possibly because he had no amygdalae, H.M. was always pretty affable around the scientists who visited him, despite never recognizing them. (Not even Milner, who worked with him for a half century.) Other amnesiacs lacked his easygoing manner, though, and a few got outright snarly. In 1992 herpes simplex—the same bug that knocked out people's ability to recognize fruits, animals, and tools—hollowed out the hippocampi and other structures inside the brain of a seventy-year-old San Diego man named E.P. He started repeating the same anecdotes over and over, verbatim, and eating up to three breakfasts each day. And despite being a former sailor who lived less than two miles from the coast, he suddenly couldn't remember even the general direction of the Pacific Ocean.

Doctors arranged to test E.P., but he grew suspicious of the "strangers"—really the same woman each time—invading his home. Every visit, he dug in his heels, and every visit, his wife had to talk him into playing nice and drag him to the kitchen table to start testing. Eventually, though, after more than a hundred visits, E.P. let his guard down. He started greeting the tester warmly, despite maintaining that he'd never seen her; he even started moving toward the kitchen table on his own to start testing. Somehow, even though his mind was telling him otherwise, his emotions remembered to trust his tester. Amnesiacs can retain negative emotional memories, too. When H.M. learned that his father had died, his conscious brain of course forgot that fact within minutes. But his emotional brain remembered, and took the news so hard that he plunged into a months-long funk, even though he couldn't explain why he felt so low. In another example, from around 1911, a Swiss doctor named

Édouard Claparède concealed a pin between his fingers before greeting a middle-aged amnesic woman; when they shook hands, he pricked her. Although she remembered nothing of this, she always withdrew her hand, and eyed him, on subsequent meetings.

Taken as a whole, this alphabetic soup of amnesiacs (q.v., e.g., H.M., K.C., E.P.) helped scientists sort out how the brain divides up responsibility* for memories. Nondeclarative memories (like motor memories) rely on the cerebellum and on certain internal clusters of gray matter such as the striatum. Episodic (personal) memories lean heavily on the hippocampus, while semantic (factual) memories utilize the parahippocampus to a much larger degree, especially for retrieval. The frontal lobes contribute as well, both in searching for memories and in double-checking that the brain has grabbed the right memory from long-term storage in the cortex. Sensory and limbic circuits also kick on to reanimate the moment in our minds. Meanwhile, the parietal and frontal lobes whisper to us that we're reviewing old information, so we don't get terrified or amorous all over again. Each step works independently, and each one can malfunction without affecting other mental faculties in the slightest.

That's the theory, at least. In reality it seems impossible to tear out any one aspect of memory—especially our episodic memories, memories of holidays and lovers and times we fell short—without tearing out so much more. K.C. knows how to play solitaire and change a tire, but he can never recall a moment of contentment, peace, loneliness, or lust. And however paradoxical it might seem, losing his past wiped out his future as well. The ultimate biological purpose of memory isn't to recall the past per se, but to prepare for the future by giving us clues about how to act in certain situations. As a result, when K.C. lost his past self, his future self died along with it. He cannot tell you what he'll do over the next hour, the next day, the next year; he cannot even imagine these things. This loss of his future self doesn't pain K.C.; he doesn't suffer or rue his fate. But in some ways that lack of suffering seems sad in and of itself. However unfair, it's hard not to see him as reduced, diminished.

In our minds, we more or less equate our identities with our memories; our very selves seem the sum total of all we've done and felt and seen. That's why we cling to our memories so hard, even to our detriment, and that's why diseases like Alzheimer's, which rob us of memories, seem so cruel. Indeed, most of us wish that we could cling to our memories more securely—they seem the only bulwark against the erosion of the self that K.C. and H.M. experienced. That's why it's such a shock to realize that the opposite burden—a hoarding, avaricious memory that *cannot* forget—can crush people's identities in the selfsame way.

~

Each morning when Moscow reporter Solomon Shereshevsky got to work, his editor assigned him and the other reporters their daily stories, telling them where to go, what to look for, and whom to interview. Despite the intricacy of the instructions, Shereshevsky never took notes, and according to some accounts he never took notes during interviews, either. He just remembered. Still, Shereshevsky wasn't a great reporter, and at one morning meeting in the mid-1920s his editor's fuse went off when he saw Shereshevsky blithely nodding at him, no pencil in hand. He called Shereshevsky out, challenging him to repeat his instructions. Shereshevsky did, verbatim—and then repeated every other word the editor had said that morning, too. When his fellow reporters stared, Shereshevsky's brow knit in confusion. Didn't everyone have complete recall? Half amazed, half creeped out, the editor sent Shereshevsky to a local neuroscientist, Aleksandr Luria.

Although a young man then, Luria had already started down the path that would make him one of the most celebrated neuroscientists of the twentieth century. He championed the romantic side of neuroscience, neuroscience that encompassed more than just cells and circuits. He wanted to capture how people actually experienced life, even the messy bits. In doing so, he swam against the current of modern science, which tends to dismiss anecdotal accounts (the plural of anecdote, after all...). But individual case studies have always been crucial to neuroscience: as with the best fiction, it's the particulars of people's lives that unveil the universal truths. Indeed, Luria's book-length case reports have been called "neurological novels," and he wrote one of his finest on Shereshevsky.

In all their years of collaboration, Luria found "no distinct limits" to Shereshevsky's memory.* The man could recite lists of thirty, fifty, seventy random words or numbers, in order, forward or backward, after hearing or reading them just once. All he needed was three seconds in between each item, to fix it in his hippocampus; after that, it was lapidary. Even more impressive, whatever he memorized stuck with him for years. In one test Luria read the opening stanzas of Dante's *Inferno* in Italian, a language Shereshevsky didn't speak. Fifteen years later, with no rehearsals in between, Shereshevsky recited the lines from memory, with all the proper accents and poetic stress. *Nel mezzo del cammin di nostra vita...*

You'd think Shereshevsky would have his pick of six-figure jobs, but like many so-called mnemonists, he drifted somewhat loserishly between careers, spending time as a musician, reporter, efficiency consultant, and vaudeville actor (memorizing lines was a snap). Unfit for anything else, he finally landed a job in what was essentially a neurological freak show, touring the country and regurgitating numbers and nonsense words to audiences. The gap between his obvious talents and his lowly status gnawed at Shereshevsky, but to Luria the discrepancy made sense. That's because Luria traced both his mnemonic prowess and his employment woes to the same root cause—excessive synesthesia.

In Shereshevsky's mind no real boundary existed between the senses. "Every sound he heard," Luria reported, "immediately produced an experience of light and color and . . . taste and touch." And unlike "normal" synesthetes, whose extra sensations are pretty vanilla (simple odors, single tones), Shereshevsky experienced full-on scenes, full mental stage productions. This became handy when memorizing items. Instead of a violet 2 or chartreuse 6, 2 became "a high-spirited woman," 6 "a man with a swollen foot." The number 87 became a stout woman cozying up to a fellow twirling his mustache. The vividness of each item made recalling it later trivial.

To then remember the *order* of such items, as in a list, Shereshevsky used a trick. He imagined walking along a road in Moscow or in his hometown (whose layout he knew by heart, needless to say) and "depositing" each image at a landmark. Each syllable of the Dante, for instance, summoned up a ballerina or goat or screaming woman, which he'd then plunk down near whatever fence, stone, or tree he happened to be passing at that moment on his mental stroll. To recall the list later, he simply retraced his route, and "picked up" the images he'd left behind. (Professional mnemonists still use this trick today.) The technique backfired only when Shereshevsky, who was rather rigid, did something foolish, like deposit images in dark alleys. In these cases he couldn't make the image out, and he'd skip the corresponding item on the list. To an outsider this seemed like a lapse, a chink in Shereshevsky's memory. Luria realized that this was actually less a failure of memory than of perception—Shereshevsky simply couldn't see the image, nothing more.

Shereshevsky's memory played other tricks as well. He could increase his pulse rate and even make himself sweat simply by remembering a time when he'd chased down a departing train. He could also (and Luria confirmed this with thermometers) raise the temperature of his right hand by remembering a time he'd held it next to a stove, while simultaneously lowering the temperature of his left hand by remembering what ice felt like. (Shereshevsky could even mentally block out pain in the dentist's chair.) Somehow his memory could override the "this is just a recollection, it's not actually happening" signal from the frontal and parietal lobes that should have quelled these somatic reactions.

Unfortunately, Shereshevsky couldn't always corral his imagination or confine it to turning mnemonic tricks. When reading a book, synesthetic images would start multiplying inside his head, crowding out the text. A few words into a story, he'd be overwhelmed. Conversations took wrong turns, too. He once asked a gal in an ice cream parlor what flavors they had. The (probably innocent) tone in which she responded "Fruit ice cream," he said, caused "whole piles of coals, of black cinders, to come bursting out of her mouth. I couldn't bring myself to buy any." He sounds insane, or like Hunter S. Thompson at his druggiest. If menus were printed sloppily, Shereshevsky's meal seemed contaminated by association. He couldn't eat mayonnaise because a certain sound *(zh)* in the Russian word for it nauseated him. No wonder he struggled to hold a job—simple instructions would mutate inside his imagination and stagger him.

Even the traveling mnemonist gig eventually became oppressive. After too many years of doing the show, Shereshevsky felt old lists of numbers and words haunting him, cacophonizing inside his skull, elbowing newer memories aside. To rid himself of them, he more or less resorted to voodoo, writing out the lists on paper and burning them. (No luck—the exorcism failed.) Relief came only from suppressing such memories, by training his mind to not acknowledge them. Only dumbing his memory down took the edge off.

Most people who met Shereshevsky considered him dim and timid, a bumbling Prufrock. Indeed, he considered himself pathetic, someone who'd wasted his talent in sideshows. But what else could he have done? With so many memories crowded into his skull—his memory actually stretched back to before his first birthday—his mind became what one observer called "a junk heap of impressions." As a result he lived in a veritable haze, nearly as befuddled and helpless as H.M. or K.C. A memory that's too good is just as broken as one that's no good at all.

To be useful, to enrich our lives, memory cannot simply record the world around us. It needs to filter, to discriminate. In fact, while we joke about a poor memory as a sieve, that's actually the wrong way around. Sieves let water leak through, but they catch substantial things—they catch what we want to preserve. In the same way, a mind functions best when we let some things, like traumatic memories, go. All normal brains are sieves, and thank goodness for that.

~

However useful, the sieve metaphor isn't perfect. Human memory doesn't just filter things. Our memories actually sculpt and rework and—with surprising regularity and slyness—distort what remains behind.

Even neuroscientists, who should know better, fall prey to distortions. Otto Loewi, whose dream about frog hearts helped prove the soup theory of neurotransmission, claimed to have had the dream over Easter weekend in 1920. But the journal in which he published his results, according to its records, received his initial submission a week before Easter that year. A few killjoy historians also think that Loewi didn't rush from his bed to the lab at 3 a.m., but instead merely wrote out the details of the experiment, step by step, then resumed snoozing. Perhaps Loewi—who loved telling tales—let the demands of narrative drama mold his memory. Similarly, William Sharpe, who harvested the glands of the giant while the family stewed in the front parlor, couldn't have done so (as he claimed) on New Year's Day, since the giant died in mid-January. Furthermore, a colleague of Sharpe's later claimed to have accompanied him on his clandestine errand—and also claimed that they picked through the giant's innards not right before the funeral but the night before, around 2 a.m. Both men cannot be correct.

Why does this happen? Why do memories get twisted like metal girders in a fire and harden into the wrong shape? Neuroscientists disagree on the answer. But one theory gaining momentum says that the very act of remembering something—which you'd think would solidify the details—is what allows mistakes to infiltrate.

When capturing a memory, neurons jury-rig a connection for the short term. They then solder those connections together with special proteins, a process called consolidation. But the brain may use those proteins for more than just capturing memories; the proteins may help retrieve and replay memories, too. Consider: If you play a beep, then shock a mouse, it sure as hell remembers this. Play the tone again, and it freezes in terror, anticipating another shock. Scientists have found, however, that they can make the mouse forget that terror. They do so by injecting a drug into the mouse's brain just before the second beep, a drug that suppresses the memory-capturing proteins. Shockingly, the *next* time the tone plays, the mouse keeps on doing mousey things. Without those proteins the memory apparently unravels, and the mouse never fears the beep again. This implies that our brains, when recalling a memory, probably don't just replay a pristine "master copy" each time. Instead, they might have to re-create and re-record the memory each time through. And if that recording gets disrupted, as it did in the mouse, the memory vanishes. This theory, called *re*consolidation, argues that there's little inherent difference between recording first mnemonic impressions and recalling them later.

Now, mice aren't little humans: humans have richer, fuller memories, and our memories work differently. But not that differently, especially on a molecular level. And if reconsolidation happens in humans—and there's evidence it does—then having to rerecord a memory each time through probably makes it labile and therefore corruptible. To be sure, we humans don't often forget events completely, like the mice did. But we do garble details,* especially personal details, all the time. As a troubling corollary, the memories that most define us—our tenderest moments, our traumas—could be most prone to distortion, since we reminisce about them most often.

So why do distortions creep in at all? Because we're human. Subsequent knowledge can always taint a memory: you can never remember your first date quite as fondly if that son of a bitch cheated on you later. So you retroactively retouch things and convince yourself that he mistreated you from the start. We also don't store memories the way computer hardware does, with each datum in a well-defined location. Human memories live in overlapping neuron circuits that can bleed together over time. (Some observers have compared this to Wikipedia editing, with each neuron able to tweak the master copy.) Perhaps most important, we feel the need to save face or goose our reputations, either by gliding over inconvenient facts or misrepresenting them. Indeed, some scientists argue that the unconscious mind confabulates—makes up plausible stories to mask our true motivations—far more often than we care to admit. Unlike victims of Korsakoff's syndrome, normal folk don't confabulate because of memory gaps. But we do tint what we recall and suppress what's convenient to suppress—until we "remember" what we want to, and can believe that a life-changing dream really did occur on Easter. Memories are memoirs, not autobiographies. And the memories we cherish most may make honest liars of us all.

CHAPTER ELEVEN

Left, Right, and Center

The largest structures in the brain are the left and right hemispheres. Human brains have striking left/right differences, especially with regard to language, the trait that best defines us as human beings.

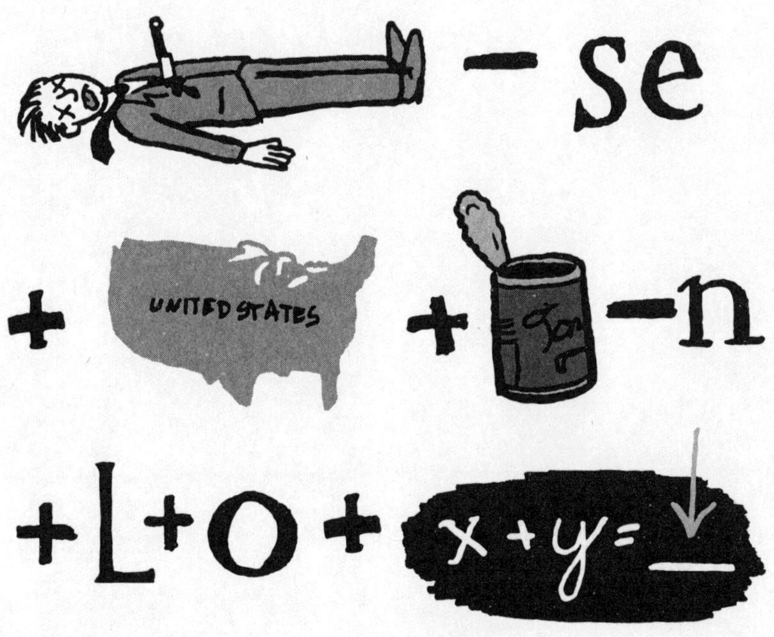

The man's name and his reasons for shooting himself—insanity? anguish? ennui?—are lost to history. But in early 1861 a Frenchman near Paris dug the business end of a pistol into his forehead and pulled the trigger. He missed. Not completely: his frontal skull bone was shattered and flipped upward like a fin. But his brain escaped unscathed. The man's doctor could in fact see the brain pulsating through the open wound—and couldn't resist reaching for a metal spatula.

Unsure whether the fellow would pass out, scream, or perhaps convulse and die, the doctor pressed the spatula down gently at various points and asked him how he felt. Although no one recorded the answer, you can imagine what the man had on his mind, so to speak. "J'ai mal à la tête, docteur. C'est—" Nothing had happened so far, but when the doctor pressed one particular spot, near the back of the frontal lobe, the man's words were snapped in two: he suddenly couldn't speak. The moment the doctor lifted the spatula, the man started up again. "Sacre bleu, doct—" The doctor pressed again, and again strangled his words. This happened over and over—each press left him sputtering, mute. The examination ended shortly thereafter, and sadly, the patient died within weeks.

A scientist named Simon Auburtin read an account of this case at a meeting of the Société d'Anthropologie in Paris on April 4, 1861. His reasons for doing so were not wholly pure. He wanted to promote the spatula-wielding doctor, a chum of his, and the case moreover supported Auburtin's pet neuroscientific theory: localization, the idea that a different region in the brain controlled each mental function. Auburtin was especially fascinated with the localization of language, an obsession he shared with his father-in-law. (The father-in-law had been cataloguing brain lesions since the 1830s, and in 1848 had bet all comers 500 francs that no one could find a widespread lesion in the frontal lobes without attendant loss of speech.) Auburtin seized on the spatula case as the best proof yet of a "language spot" inside the brain.

Believing in localization did not put Auburtin in the majority among his colleagues, most of whom disdained localization, dismissing it as phrenology 2.0. The original phrenology movement had died in ridicule decades before, and Auburtin himself conceded that phrenologists had gone overboard in tracing things like atheism or a "carnivorous instinct" to specific head bumps; he wanted to salvage only the general principle of brain specialization. But no matter how carefully Auburtin couched his ideas, they retained the stink of quackery. It didn't help that localization violated many scientists' metaphysical beliefs about the brain and soul being indivisible into smaller units. As you can imagine, this wasn't the kind of argument you could settle in an hour, and the meeting that day in April deteriorated into squabbling.

In the audience that afternoon, taking notes for the house newsletter, sat the society's thirty-seven-year-old secretary, Paul Broca. The son of an army surgeon, Broca had come to Paris a dozen years before. At first he'd whiled away his days writing and painting; he later found a teaching job but detested it, and was so hard up that he considered striking out for America. By his late twenties he'd righted himself and found work as an anatomist and surgeon. But with every passing year he devoted more and more time to his life's passion—skulls, which he amassed a huge collection of. More generally, Broca loved anthropology, and he'd cofounded the Société d'Anthropologie in 1859. He'd envisioned free-ranging discussions about human origins and primitive societies (and skulls), not quibbles about brain localization. Indeed, the topic held little interest for him—at least until he met Tan.

Tan's real name was Leborgne. An epileptic since childhood, Leborgne had earned his living making hat lasts, the wooden molds around which milliners sculpted their chapeaus. But years of epileptic damage eroded his ability to speak, and by age thirty-one all he could say, in response to any question, was "Tan tan." That soon became his nickname, and in 1840, unfit for anything else, Tan was committed to the Bicêtre, a half hospital, half nursing home, outside Paris. He did not respond well to this confinement. Perhaps the frustration of being inarticulate overwhelmed him, or perhaps, as with H.M., other patients tormented him. Regardless, Tan turned into a real prick after being committed. Other Bicêtre patients found him egotistical, mean, and vindictive; some accused him of thievery. The strange thing was that, when pushed too far, Tan could say something besides "Tan tan." He'd scream *Sacré nom de Dieu!* in their faces, scandalizing everyone within earshot. But Tan couldn't swear voluntarily, only in the throes of rage.

However vicious, Tan didn't deserve what happened next. In 1850 he lost all feeling in his right arm; four years later his right leg became paralyzed, and he spent the next seven years confined to bed. In those days bedsores frequently turned lethal, and because Tan never soiled his sheets, nurses rarely changed his linens or rotated him. He also had no feeling on his right side, so by the time someone noticed the gangrene, it had chewed up his right leg from heel to buttock. He needed an amputation, and on April 12, 1861, his doctors presented him to a newly hired surgeon at Bicêtre, Paul Broca.

Broca started by taking Tan's clinical history. Your name, monsieur? "Tan." Occupation? "*Tan* tan." The nature of your troubles? "Tan—tan!" Each "tan" came out pure and sweet and dulcet—Tan's voice still sounded nice—but the absurdist dialogue meant nothing to Broca. Thankfully, Tan had become a master mime and could communicate through hand signals. For instance, when Broca asked how long he'd been in Bicêtre, Tan flashed jazz hands with the fingers of his left hand four times, then his index finger once—twenty-one years, the right answer. To check whether this was a lucky guess, Broca asked the same question the next day. Just to be sure, Broca asked a third time, the following day. At this, Tan realized he was being tested and screamed, *"Sacré nom de Dieu!"* (When reporting this curse in his case report, Broca used euphemistic dashes.) Broca determined from these interviews that Tan, despite losing the ability to speak, could still understand language.

Per his duty, Broca amputated Tan's leg. But the gangrene had weakened Tan too much, and he died the morning of April 17. Within twenty-four hours Broca, still mulling over the recent Société debate about the "language spot," opened Tan's skull.

Inside Broca found a mess. The left hemisphere looked deflated, and it almost disintegrated when touched. The frontal lobe especially looked nasty: it contained a rotten cavity "the size of an egg," with yellow ichor pooled inside. Despite the mess, Broca's trained eye noticed a crucial detail: that the putrefaction, while widespread, seemed to get worse the closer he got to a central point. And the bull's-eye of this putridity sat near the back of the frontal lobe—exactly where the doctor in the case report had pressed the spatula. Broca deduced that this was the original lesion. And because Tan's original symptom had been loss of speech, Broca concluded that this area must be a language node. In deciding this, Broca effectively threw his chapeau in with Auburtin and the neophrenologists, a dicey career move. Even more risky, Broca decided to present Tan's brain to his beloved Société at its next meeting—April 18, that very afternoon.

The meeting had all the makings of high scientific drama. Broca entered with Tan's freshly disembodied brain—confronting a skeptical audience, yet armed with the first solid evidence, ever, of brain localization. It might have been Huxley versus Wilberforce part deux, and indeed, Broca's latter-day disciples have infused his talk that day with an almost supernatural significance. In truth, Broca did little more than present the brain for inspection and summarize Tan's medical history; he mentioned his conclusion about the language node only briefly, without pressing it. His would-be opponents all but yawned, and as soon as Broca finished they delved into a much juicier debate about race, brain size, and intelligence.

Race, brain size, and intelligence obsessed Broca, too, and he had much to contribute to the discussion, which ended up dominating the society's agenda for months. Nevertheless, Broca kept mentioning Tan's brain here and there at subsequent meetings, and one cog in his own brain kept turning over Tan's "aphasia," as the neurological loss of speech is now known. He preserved Tan's brain in alcohol, then placed the pickled mass in a jar for future study. Meanwhile he scouted around for other aphasics, and soon encountered a patient who deserves to be every bit as famous as Tan.

Like Tan, "Lelo" earned his nickname based on what little he could say. An octogenarian ditchdigger, Monsieur Lelong had suffered a stroke eighteen months before Broca met him in October 1861. He'd lost all ability to speak, save for five words: "Lelo," his name for himself; *"oui"; "non";* "tois," for *trois,* three, which stood for all numbers; and *"toujours,"* always, which stood for the rest of the dictionary. Ask him how many daughters he'd sired, and he'd say "tois" and hold up two fingers. Ask him what he did for a living, and he'd say "toujours," and mime shoveling dirt.

History records little else about Lelo, except that complications from a broken femur soon killed him. But when he died, Broca performed probably the most important brain autopsy since Henri II's. Going in, Broca felt anxious, edgy: if he found no lesion in Lelo's brain—or a lesion in the wrong place—he'd be mocked. He sawed the skull carefully and cracked open the shell. He needn't have worried. Whereas Tan's brain looked pulverized, with widespread putrefaction, Lelong's brain had a single BB hole of damage. And Broca himself must have cried *Sacré nom de Dieu!* upon seeing the location: near the back of the frontal lobe. This location* is now known as Broca's area.

Broca's announcement of a language node inside the human brain caused no big stir among the public. (Paris newspapers were tittering instead about the underwhelming premiere—full of catcalls and raspberries—of Richard Wagner's *Tannhäuser.*) But the discovery ricocheted through the learned societies of Europe, leaving scientists agog. *Could localization be real?* Two subsequent developments suggested yes. First, Broca confirmed his initial findings in more patients. After 1861 doctors started referring aphasics to Broca for further study, and by 1864 he'd done autopsies on twenty-five of them. Every victim save one had a lesion in the rear frontal lobe. Moreover, the nature of the damage — tumors, strokes, syphilis, trauma — didn't matter, only its location, location, location.

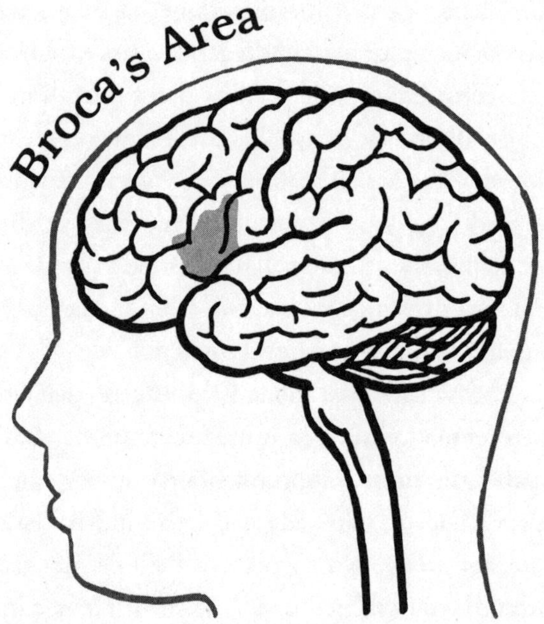

The second development had even more profound consequences for understanding how language works inside the brain. In 1876 a twenty-six-year-old German medical student named Karl Wernicke (of Wernicke-Korsakoff fame) discovered a new type of aphasia. Specifically, Wernicke found that lesions near the back of the temporal lobe — well distant from Broca's area — destroyed the *meaning* of language for people. Whereas Broca's aphasics knew what they wanted to say but sputtered in saying it, Wernicke's aphasics could string together sentences of Proustian length, with quite fetching rhythms; the sentences just didn't make sense. (Some neuroscientists call this a word salad—random chunks of phrases tossed together. I'd call it *Finnegans Wake* syndrome.) And unlike Broca-type aphasics, who get quite frustrated, Wernicke-type aphasics remain oblivious; doctors can spout gibberish right back at them, and they'll nod and grin. Generally speaking, a broken Broca's area knocks out speech production, while a wrecked Wernicke's area impairs speech comprehension.

Functionally, Broca's area helps the mouth form and articulate words, so when this area falters, sentences get choppy and people have to pause frequently. Moreover, it helps generate proper syntax, so Broca aphasics use almost no syntax or conjunctions to string concepts together: "Dog—bit—girl." Wernicke's area, in contrast, links words to their meanings—it fuses signifier with signified inside your brain. To see how the areas work together, imagine that the person next to you suddenly says "Zeppelin." First your ear relays this input to your auditory cortex, which in turn relays it to Wernicke's area. Wernicke's then dredges up the proper associations in your memory, causing you to glance skyward, hear a guitar riff, or think, "Oh, the humanity!" Sound and meaning are thereby united. If you decide to repeat "zeppelin" aloud (and why not?), Wernicke's area first matches up the concept of "zeppelin" with the auditory representation stored in your brain. Wernicke's then sends out a signal that arouses Broca's area, which in turn arouses the strip of motor cortex that controls your lips and tongue. If your Wernicke's cannot match up the words and ideas, it's word salad time. (Infants cannot produce or understand language in part because their Wernicke's area hasn't matured.) If Broca's area fouls up, you sputter.

Beyond finding a new language node, Wernicke made a more general point about language inside the brain, a point worth italicizing: *there is no single "language spot" up there.* As with memory, many different regions contribute to understanding and producing language, which explains why people can lose the ability to speak without losing the ability to comprehend, or vice versa. If other language nodes crumble, or if the white-matter cables between two language nodes get severed, language skills can break down in other ways as well, some of them startlingly specific.

Some stroke victims can remember nouns but not verbs, or vice versa. People fluent in two languages can lose either one after trauma, since first and second languages* draw on distinct neural circuits. Language deficits can even interfere with math. We seem to have a natural "number circuit" in the parietal lobe that handles comparisons and magnitudes—the basis of most arithmetic. But we learn some things (like the times tables) linguistically, by rote memorization. So if language goes kaput, so too will those linguistically based skills.

More strikingly, some people who struggle to string even three words together can sing just fine. For whatever reason, melody and rhythm can bypass broken circuits and jump-start language production—allowing someone who stammers through "I—like—ham" to rip through "The Battle Hymn of the Republic" moments later. (After being shot in the brain, former congresswoman Gabrielle Giffords learned how to speak again by practicing with song lyrics, including "Girls Just Wanna Have Fun.") Similarly, emotions can also resurrect dead language circuits: many aphasics (like Tan) can swear if provoked, but never intentionally. The dissociations between singing, speaking, and swearing imply, again, that our brains don't have a single language spot; there's no neurological "pantry" where we keep our words.

Perhaps the most startling example of language disconnect is called alexia sine agraphia, a reading disorder. Reading actually requires a higher degree of neurological dexterity than speaking does. Printed words enter our brains through the visual cortex easily enough, but because we humans started reading so late in our evolutionary history—around 3000 BC—the visual cortex doesn't naturally wire itself up to Wernicke's area. (Why would it?) A little training with *Dick and Jane* can nevertheless rewire the brain and patch those two areas together—allowing us to conjure up concepts and stories from mere dashes of ink. Reading changes the way our brains work.

People with alexia sine agraphia, however, cannot read a lick, because of broken axons in the visual cortex: the curves and shapes of letters get into their brains just fine, but the data never reach Wernicke's area and never get converted into meaningful information. As a result, sentences look like they're written in цырилиц or 中国的. Yet these people can write just fine, because the brain's meaning centers can still access downstream motor circuits that control handwriting. This leads to the farcical situation of someone being able to write down a sentence—"I'm allergic to beer"—but not being able to read what he just wrote.

As much as anything, language makes us human, and Broca earned his bust on the Mount Rushmore of modern neuroscience largely for discovering the first language node. Truth be told, though, Wernicke's idea of language circuits is more in harmony with our current understanding of language. And while Broca also usually gets credit for discovering brain localization, Auburtin* and even the phrenologists pushed the idea of localization first, and pushed it harder. It was simply Broca's eminence, his vivid clinical reports, and especially his luck in finding Tan and Lelo that transformed those other scientists' intuitions into scientific fact.

~

By all rights Broca should also share credit for the other major discovery usually attributed to him, brain lateralization. By the mid-1800s scientists knew that the left hemisphere controls the right side of the body and vice versa. But scientists still believed, deeply, in brain symmetry—the idea that both halves of the brain worked the same way. After all, each hemisphere looked identical, and in no other paired body part (eyes, kidneys, gonads) did lefty and righty function differently. So when doing autopsies on aphasics, Broca ignored any hemispherical differences and concentrated solely on longitude and latitude. Only in early 1863 did he realize that all his aphasics so far had *left* frontal lobe damage. He chewed on the potential meaning of this in private—could the left hemisphere control language? "But I could not easily resign myself," he later admitted, "to such a subversive consequence."

Others proved less timid. In March 1863, while Broca hemmed and hawed, an obscure country physician named Gustave Dax submitted a three-decades-old manuscript to the Académie Nationale de Médecine in Paris, hoping to publish it. In an accompanying letter Dax explained that the manuscript belonged to his late father, Dr. Marc Dax, who had compiled case reports on dozens of patients who'd lost the ability to speak after suffering frontal lobe damage. Dax *père* had then presented the manuscript at a conference in Montpellier in 1836, but had been unjustly ignored ever since. Because all his patients had injuries in roughly the same place, the elder Dax concluded that the frontal lobe contained a language spot—exactly what Broca had proposed only two years before. Furthermore, since all these lesions appeared on the left side, the left hemisphere must control language—exactly the idea Broca was playing around with now.

The history of science is full of examples of two or more people discovering something independently—oxygen, sunspots, calculus, the periodic table. But few priority disputes have proved as messy as the Broca-Dax affair. Broca made his first tentative public statement about the left hemisphere being the seat of language in early April 1863, just days after the Dax manuscript surfaced in Paris. The contents of a manuscript submitted to the Académie were supposedly confidential, but Broca had friends there, and almost certainly knew of its conclusions ahead of time. What's more, the Académie took its sweet time reviewing the paper for publication, first referring it to committee (that ultimate tool of bureaucratic obstruction) and then holding it for over a year. Dax eventually had to publish the manuscript himself, and the delay gave Broca time to develop his ideas.

However, the younger Dax—by all accounts an obnoxious fellow—didn't take this subterfuge lying down. He railed against the Académie's stall tactics and rallied support among scientists in southern France, who generally resented their snooty Parisian colleagues. Dax also accused Broca of stealing his dear father's ideas, by purposely failing to cite his father's work. Broca took this charge seriously and began to hunt down other scientists who'd attended the Montpellier conference in 1836, to ask about Dax's presentation there.

Oddly, though, none of the attendees remembered Dax's work. And after a few months of dead ends, Broca came away unsure whether Dax had even attended the conference, much less presented there. As a matter of fact, the only evidence that the elder Dax had ever studied language lesions was the original draft of the manuscript, which supposedly dated from the 1830s. That provenance, however, depended on the younger Dax's word, and Broca naturally grew suspicious. He even analyzed the writing style of both Daxes, to see if Dax *fils* had tried to pass off a forgery. (Broca ruled the document authentic, but he was no linguist.)

The Broca-Dax affair remains clouded today. There's no question Broca was the superior scientist. Like Darwin with natural selection or Mendeleev with the periodic table, Broca didn't discover lateralization alone; but also like those men, his work was an order of magnitude more developed than any rival claim. Dax didn't even confirm the location of his patient's lesions with autopsies; he simply guessed, based on where patients said they'd been smacked. Nevertheless, Dax got it right—the left brain does control language—and in science, getting it right first often counts for everything.

The tougher debate is how much Broca knew and when he knew it. The younger Dax's bitching notwithstanding, Broca almost certainly didn't rip off his father wholesale. But did the manuscript influence Broca? Perhaps it was a coincidence that Broca felt confident enough to start speaking about left-side specialization shortly after the manuscript arrived in Paris. Or perhaps hearing of the manuscript convinced Broca that he was on the right track. Most historians agree that Marc Dax and Paul Broca probably did discover left-right lateralization independently. But good luck determining how much Dax influenced Broca, or whether Dax gave him the courage to pursue an avenue he might not have.

Amid all this squabbling, Broca withdrew somewhat from neuroscience, and after about 1866 he decided to focus more on other scientific topics, such as skulls. In 1867 he wowed the world by determining that a pre-Columbian skull from Peru—which had a square hole carved into it—was evidence of ancient neurosurgery. Broca even declared, correctly, that the patient had survived the operation, based on healing scars around the hole's rim. Around the same time he saved a man's life by performing the first neurosurgery based on localization theory. A patient had lost the ability to speak after head trauma, and instead of removing half the man's skull to explore, Broca opened a small hole over his eponymous area and relieved the pressure.

Meanwhile, Broca began dabbling in politics. During a tumultuous coup attempt in 1871, he smuggled 75 million francs' worth of gold to Versailles in a hay cart (some sources say a potato cart), to help the exiled government. The powers that be never rewarded him, but the French people did elect Broca a "lifetime Senator" in 1880. Before he could really enjoy the honor, though, he died a few months later at age fifty-six—fittingly, of brain trouble, a hemorrhage.

After his untimely death scientists more or less beatified Broca, and brain lateralization became a pillar of twentieth-century neuroscience. In fact, as so often happens, this former heresy became the new orthodoxy: by the 1950s most neuroscientists had declared the left hemisphere home to not just language but all of our highest faculties and skills. Humankind *was* its left brain. And not content to merely praise the left brain, scientists simultaneously demeaned the right brain, dismissing it as the left's slower, imbecilic, even "retarded" twin. It would take one pissed-off Nazi, and decades of follow-up work, to prove otherwise.

~

In 1944 a thirty-year-old American officer, W.J., leapt out of a plane over Holland to help liberate the Dutch. His parachute opened only partway, and he hit the ground like a sandbag, breaking his leg and knocking himself unconscious. Upon waking he began pissing blood, and soon became a Nazi captive. At some point—perhaps while being herded around a POW camp—he enraged a guard, who wound up and cracked him on the skull with his rifle butt. W.J. crumpled, and probably suffered a brain hemorrhage. He barely got any treatment over the next year, and lost almost a hundred pounds.

After the war W.J. found work as a payroll courier in Los Angeles. But he started having what are called "absences": he'd start his car, pull out—then find himself fifty miles away, having no idea how he'd got there. He began having seizures, too. His aura felt like a Ferris wheel rumbling to life inside him, and his head would jerk left; he'd grimace and occasionally yell "Bail out, Jerry!" before collapsing. He didn't soil himself, but he banged and scraped his head a lot and once fell into a fire. Perhaps worst, the frequency of seizures—up to twenty per day by the later 1950s—left him mentally dazed. Whereas before the war he'd read Greek history and Victor Hugo with enthusiasm, now he could manage only newspaper headlines. So in 1962 he agreed to let two L.A. surgeons perform an operation every bit as desperate as H.M.'s surgery a decade earlier. They proposed slicing clean through W.J.'s corpus callosum.

You can't see the corpus callosum unless you peel the brain's two halves apart and peer into the gully. It looks like a bundle of off-white twine, and it connects the two hemispheres like Siamese twins. It's one of the few brain structures we have just one of, and in past centuries at least a few scientists pinned the indivisible human soul there for this reason. By the 1900s scientists didn't view the corpus callosum as a sanctum sanctorum anymore, but damn them if they knew what it did do. It consists of 200 million white-matter fibers, which implied a role in interhemispheric communication. (The next beefiest bundle connecting the hemispheres contains just 50 thousand fibers.) Still, X-rays showed that some people were born without a corpus callosum, and they seemed fine.

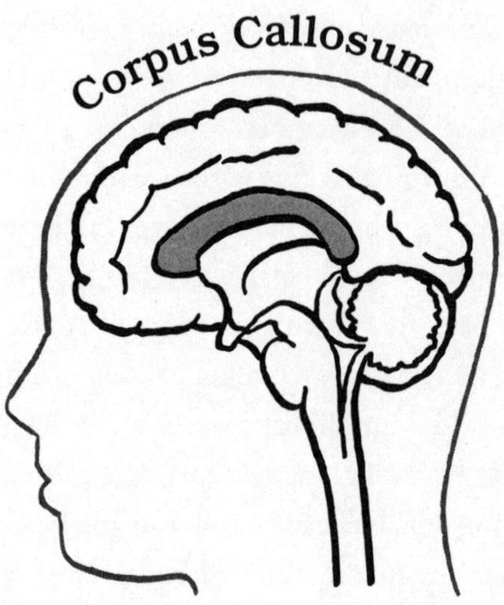

Neuroscientists could name only one thing the corpus callosum definitely did: spread seizures. By using a helmet, they could monitor the electrical patterns of a seizure within the brain. For whatever reason, small epileptic storms seemed to gather momentum after reaching the corpus callosum, and would soon overrun the whole globe. That danger did suggest a way to head off seizures, though—sever the corpus callosum. The two L.A. surgeons began practicing this procedure on cadavers, and they finally convinced W.J. to submit in 1962. They drilled two holes in his skull, one forward, one aft, then slid spatulas inside to hoist up his lobes. You might think this type of surgery would go quickly—just stick a knife inside and start carving—but it actually required ten hours of work: whereas upper brain tissue can be scooped like tapioca, the corpus callosum is tough as gristle.

Recovery was slow, but W.J. started talking a month later and walking three months after that; doctors also monitored his fine motor skills and rejoiced to see him do coordinated, bimanual tasks like lighting cigarettes. (It was a different era.) Best of all, W.J.'s seizures vanished. The surgery had aimed to confine his fits to one hemisphere, but for unknown reasons it all but eradicated them. For the first time in a decade he started sleeping through the night, and he gained forty much-needed pounds. Just as important, W.J. didn't suffer any H.M.-like crisis: his personality and speech and memories remained intact. Encouraged, the L.A. surgeons started performing more callosectomies. And aside from the short-term pains of surgery—one patient woke up, he quipped, with a "splitting headache"—patients showed no ill effects. They could still read, reason, and remember; they could talk, walk, and emote. Their minds worked exactly the same as before.

And yet—could that possibly be true? Could hacking through 200 million fibers produce zero side effects? Neuroscientist Roger Sperry didn't buy it, and set out to prove otherwise.

Proving otherwise was a habit with Sperry. He had an atypical background for a scientist, having focused as much on sports as academics when young. He set the state javelin record in high school in Connecticut and lettered in baseball, basketball, and track at Oberlin College. When not training, he spent his undergraduate hours mooning over seventeenth-century poetry. But his Psych 101 class intrigued him, and after he hung up his jock at Oberlin he stuck around and got a master's degree in psychology. He then pursued a doctorate in zoology at the University of Chicago, where—in an impolitic but no doubt satisfying move—he demolished his thesis advisor's lifework.

Paul Weiss promoted the trendy "blank slate" theory of brain function. He argued that any neuron could do the job of any other neuron, and that brain circuits could be rewired to an infinite degree. Sperry thought this superplasticity sounded pretty cool, so in 1941 he started a series of diabolical experiments on rats to test the idea. These experiments involved, among other things, opening up the rats' two back legs, finding the nerves that carried pain signals to the brain, and switching them, so that the left pain nerve was now located in the right leg, and vice versa.

Once a rat recovered from surgery, Sperry placed it on an electrified grille where, if it stepped on a certain spot, it got a shock. The result was black comedy. If the rat shocked its back left leg, its brain (thanks to the switched nerves) felt the sting of pain in the right leg. So the rat jerked the right leg up and began limping. Unfortunately, this put more weight on the left leg, which is where the wound actually was. Worse, as soon as the rat wandered near that electrified spot again, its left leg got another shock. Which because of the switched nerves made it seem like the right leg was hurting even worse. Which made the rat favor the damaged left leg all the more. Which led to more pain and more shocks the next time around, and so on, a vicious cycle. Crucially, and contra Weiss, the rat's neurons never learned better. Month after month passed, but no matter how many times the poor booby shocked one leg, it always hiked the other one up.

Sperry did even more Dr. Moreauvian things to fish. He'd pop out their eyeballs, sever the optic nerves, rotate the eyeballs 180 degrees in the socket, then sew them back in. Fish nerves can regenerate themselves no problem, so the fish learned to see again. But because the eyeballs had been rotated, the nerves rewired themselves backward—forcing the fish to see the world upside down. Wriggle a worm below its jaw, and it snapped upward; dangle a morsel above, and it lurched downward. And once again, the fish never ever unlearned this behavior.

Sperry determined from the rat and fish stunts that all creatures have some hardwired neural circuits: certain neurons are born to do certain jobs and cannot learn other tasks. That's not to say there's no plasticity within the brain (especially in humans). But Sperry demolished the idea that we're born with blank neurological slates.

Not satisfied with ruining Weiss's work, Sperry took a postdoc at Harvard University and slew his advisor there, too. Starting in the 1920s Karl Lashley had helped popularize that all-time-classic psychology experiment, running rats through mazes. In his case, after the rats had learned a maze Lashley would anesthetize them, make lesions in their brains, then test them again. To his shock, no matter where he inflicted damage, the rats could usually still negotiate the labyrinth, provided he didn't damage too much tissue overall. In other words, he claimed, the location of the lesions didn't matter, only their size. Based on this work Lashley developed a theory of antilocalization. He admitted that brains must have some specialized components. But for advanced tasks, such as learning mazes, Lashley argued that creatures utilized all parts of the brain simultaneously. As a corollary, Lashley promoted the idea, prominent before H.M., that all parts of the brain contribute equally to forming and storing memories.

For Lashley's theory to work, distant regions in the brain—even regions not connected by axon wires—had to communicate almost instantly. So he downplayed the idea that neurons sent messages only to direct neighbors, bucket-brigade-style. Instead, he envisioned neurons emitting long-distance electrical waves—similar to the then-new medium of radio. Sperry once again thought this sounded awesome, but he once again proved too good a scientist. Starting in the mid-1940s, he opened up the skulls of cats and inlaid their brains either with strips of mica (to insulate) or with tantalum wires (to short-circuit). Either addition should have disrupted the electrical waves propagating through the brain and thereby shut down higher thinking. Nope. Sperry ran the felines through every neurological test he knew, and they acted exactly the way cats always have and always will. This killed Lashley's theory* of long-distance electrical communication and reinforced the belief in neuron-to-neuron chemical communication.

To the relief of advisors everywhere, Sperry opened up his own lab at the California Institute of Technology in 1954. After settling in, he decided to expand some earlier work he'd done on the corpus callosum. These experiments involved slicing through this bundle in cats and monkeys and monitoring their behavior.

All in all, the split-brain animals seemed normal—at least mostly.

Every so often they'd do something funny, something off. For instance, if he taught a split-brain cat to navigate a maze while wearing an eye patch, then switched the patch to the other eye and put the cat back in the maze, it would start getting lost again.* That didn't happen to controls. Sperry saw enough quirks like this to doubt that humans who had their corpus callosums severed would escape with no side effects. So when the L.A. surgeons asked Sperry to test W.J. and other callosectomy patients, Sperry agreed—and once again proved otherwise.

The tests—run by Sperry and his graduate student, Michael Gazzaniga—took place in three-hour blocks each week. At first W.J. seemed normal. He handled everyday interactions just fine, and even extensive psychological testing turned up no oddities. Then came the tachistoscope. The tachistoscope was basically a mechanical shutter hooked up to a projector. It snapped open and shut quickly, allowing scientists to flash images onto a screen for a tenth of a second. Before the 1950s, the tachistoscope was best known for helping train American fighter pilots during World War II. Psychologists would flash silhouettes of planes—both Allied and enemy aircraft—to the flyboys, who, after suitable training, learned to distinguish good guys from bad guys in an instant.

Rather than planes, W.J. saw flashes of words or objects. He sat at a table six feet away from a white screen and fixed his gaze on the center. Beneath the table sat a telegraph key, which he pressed to signal that he'd seen an image. *Tap.* After each round, as a backup, Sperry and Gazzaniga also asked W.J. to state whether he'd seen any image, yes or no. The crucial aspect of the experiment was this: Sperry and Gazzaniga would flash the word or object only on one side of the screen—far to the left or far to the right of the center line. As a result, the image entered only one side of W.J.'s brain. His response to these fleeting images made the scientists' skin tingle.

With images flashed to his right side, W.J. responded as you'd expect. Those images entered his left brain, which controls both language and his right hand. So his right hand pressed the telegraph key and he answered yes, he'd seen an image. Images flashed to his left side were a different story. These images entered his right brain, which can't produce language; nor could the right brain signal the left brain to speak up, because of the split corpus callosum. So W.J. denied seeing anything. But his left hand still pressed the telegraph key. His left hand *knew,* even if his left brain didn't. This happened over and over. W.J. would insist he'd seen nothing—nothing—nothing. Meanwhile he was practically tapping Morse code beneath the table.

Other split-brain patients showed a similar disconnect between right and left. In one test, Sperry and Gazzaniga blindfolded patients and placed pencils, cigarettes, hats, pistols, and other objects in their left palms. The patients could use these objects just fine—scribble, puff, doff the hat, pull the trigger—but they could never name them. In another test the scientists used the tachistoscope to flash "hot" and "dog" simultaneously on opposite sides of the screen, then asked people to draw a picture* of what they'd seen. When normal people took this test, they drew a Nathan's Famous—wiener, bun, maybe some mustard. The split-brain people drew two pictures: a pooch with their right hand and a withering sun with their left. (They also botched head/stone and sky/scraper.) In sum, they failed any test that required the right brain and left brain to share information. Minus a corpus callosum, each hemisphere remained isolated.

Sperry and Gazzaniga didn't just scour for deficits, though. Split-brain patients also helped them tease out the unique talents of each hemisphere—what we now call left brain versus right brain thinking. Again, scientists at the time considered the left brain superior in pretty much every skill that mattered. But split-brain patients revealed that the right brain recognized faces better: when split-brain people saw an Arcimboldo portrait, the left brain saw the component fruits and vegetables, while the right brain saw the "person." The right brain also did a better job at spatial tasks such as rotating objects mentally, or determining how big a circle was after seeing a small arc. Perhaps

A "portrait" by Giuseppe Arcimboldo. Depending on the brain damage, some victims see only the constituent fruits and vegetables, some see only the overall face. The right brain also tends to notice the face alone, while the left zeroes in on the comestibles.

most interesting, the right brain outgambled the smarty-pants left brain. Imagine a game in which you draw marbles from a giant tub. Eighty percent are blue marbles, 20 percent red, and if you guess the right color before each draw, you get a dollar. In tasks like this, whole-brain people generally guessed blue 80 percent of the time, red 20 percent—a moronic strategy. If you do the math, you'll get only 68 percent correct this way. Better to guess blue every single time, since you're guaranteed 80 percent success. Rats and goldfish (in animal-appropriate versions of the game) got this right: they always guessed the same color. The left brain of split-brain people guessed like normal people. The right brain didn't. It guessed the way rats and goldfish did, and cashed in.

Building on this work, other neuroscientists discovered other right-brain talents, further eroding the left brain's hegemony. The right brain proved a better musician in split-brain folks, and its superior spatial skills allowed it to read maps more fluently. The right brain even dominated certain aspects of language. If the right-brain equivalent of Broca's speech area gets damaged, people end up with a condition called aprosodia. They understand the literal meaning of words but remain oblivious to the rhythms and emotional nuances of actual conversation—the things that make language sizzle. The right brain tends to dominate anything we think of as "arty." In fact, if the domineering left brain suffers damage, the right brain's artistic instincts often come to the fore. There are well-documented cases of people suffering left-brain trauma and suddenly becoming obsessed with painting or poetry, things they never cared a whistle about before. Similarly, many idiot savants suffered prenatal left-brain damage, and their uncanny talents (like musical mimicry) might actually be normal right-brain talents that found an outlet.

Nevertheless, despite these distinct talents, Sperry and Gazzaniga cautioned against making too much of right-brain/left-brain differences. It's not as if one hemisphere speaks or paints all by itself while the other just sits there, twiddling its axons. The left-brain/right-brain relationship is more complementary, more like the relationship between the left and right hands. Most people have a dominant right hand, but the left still helps tie shoes, type, pour drinks, and scratch certain places. In the same way, the brain cannot complete most tasks without both hemispheres working in concert. One great example here is scientific reasoning. Split-brain patients have demonstrated that the right brain does a better job determining whether two events are causally linked (that is, determining whether A actually caused B, or whether the connection was spurious); it also keeps a better, more faithful record of what we see, hear, and feel. The left brain does a superior job culling patterns from data, and only it can take basic information and leap to something new, to a law or principle. In sum, both sides of the brain sense reality but do so in different ways, and without their unique perspectives, we'd have gaps in our scientific understanding.

Scientists suspect that left-right specialization first evolved many millions of years ago, since many other animals show subtle hemispheric differences:* they prefer to use one claw or paw to eat, for instance, or they strike at prey more often in one direction than another. Before this time, the left brain and right brain probably monitored sensory data and recorded details about the world to an equal degree. But there's no good reason for both hemispheres to do the same basic job, not if the corpus callosum can transmit data between them. So the brain eliminated the redundancy, and the left brain took on new tasks. This process accelerated in human beings, and we humans show far greater left/right differences than any other animal.

In the course of its evolution the left brain also took on the crucial role of master interpreter. Neuroscientists have long debated whether split-brain folks have two independent minds running in parallel inside their skulls. That sounds spooky, but some evidence suggests yes. For example, split-brain people have little trouble drawing two different geometric figures (like ⊔ and ⊏) at the same time, one with each hand. Normal people bomb this test. (Try it, and you'll see how mind-bendingly hard it is.) Some neuroscientists scoff at these anecdotes, saying the claims for two separate minds are exaggerated. But one thing is certain: two minds or no, split-brain people *feel* mentally unified; they never feel the two hemispheres fighting for control, or feel their consciousness flipping back and forth. That's because one hemisphere, usually the left, takes charge. And many neuroscientists argue that the same thing happens in normal brains. One hemisphere probably always dominates the mind, a role that Michael Gazzaniga called the interpreter. (Per George W. Bush, you could also call it "the decider.")

Normally, having an interpreter/decider benefits people: we avoid cognitive dissonance. But in split-brain patients, the know-it-allness of the left brain can skew their thinking. In one famous experiment Gazzaniga flashed two pictures to a split-brain teenager named P.S.—a snowscape to his right brain and a chicken claw to his left brain. Next, Gazzaniga showed P.S. an array of objects and had him pick two. P.S.'s left hand grabbed a snow shovel, his right hand a rubber chicken. So far, so good. Gazzaniga then asked him why he'd picked those things. P.S.'s linguistic left brain knew all about the chicken, of course, but remained ignorant of the snowscape. And, unable to accept that it might not know something, his left-brain interpreter devised its own reason. "That's simple," P.S. said. "The chicken claw goes with the chicken, and you need a shovel to clean out the chicken shed." He was completely convinced of the truth of what he'd said. Less euphemistically, you could call the left-brain interpreter a part-time confabulator.

Split-brain patients confabulate in other circumstances, too. As we've seen, thoughts and sensory data cannot cross over from the left hemisphere to the right hemisphere, or vice versa. But it turns out that raw emotions can cross over: emotions are more primitive, and can bypass the corpus callosum by taking an ancient back alley in the temporal lobe. In one experiment scientists flashed a picture of Hitler to a split-brain woman's left side. Her right brain got upset and (the right brain being dominant for emotions) imposed this discomfort onto her left brain. But her linguistic left brain hadn't seen Hitler, so when asked why she seemed upset, she confabulated: "I was thinking about a time when someone made me angry." This trick works with pictures of funeral corteges and smiley faces and Playboy bunnies, too: people frown, beam, or titter, then point to some nearby object or claim that some old memory bubbled up. This result seems to reverse neurological cause and effect, since the emotion came first and the conscious brain had to scramble to explain it. Makes you wonder how much we actually grasp about our emotions in everyday life.

Along those lines, split-brain people can help illuminate certain emotional struggles we face. Consider P.S., the teenager who confabulated about chickens and shovels. In another experiment scientists flashed "girlfriend" to his right hemisphere. In classic split-brain fashion, he claimed he saw nothing; but in classic teenage fashion, he giggled and blushed. His left hand then used some nearby Scrabble tiles to spell L-I-Z. When asked why he'd done that, he said he didn't know. He certainly wouldn't do anything as stupid as like a girl. Tests also revealed conflicting desires in his right and left brain. P.S. attended a fancy finishing school in Vermont, and when asked what he wanted to do for a living, his left brain bid him say "Draftsman," a respectable career. Meanwhile his left hand spelled out "automobile race[r]" with tiles. His brain even betrayed a red/blue political divide: post-Watergate, his left brain expressed sympathy for President Nixon, while his right brain hinted it was glad to see Tricky Dick go. When facing a crisis or controversy, we often talk about feeling torn or being of two minds. Perhaps those aren't just metaphors.*

~

This left-right asymmetry within the brain affects how we read emotions in other people as well. Imagine simple line drawings of two half-smiley, half-frowny faces, one with the smile on the left side of the face, one with the frown on the left. In a literal sense, these faces are equal parts sad and happy. But to most people the emotion on the left side (from the viewer's point of view) dominates, and determines the overall emotional tenor. That's because whatever's in your left visual field taps into the emotion-dominant and face-dominant right brain. Along those lines, if you bisect a person's photograph and view each half independently, people usually think he "looks like" the left half more than the right half.

Artists have long exploited this left-right asymmetry to make their portraits more dynamic. Generally, the left half of someone's face (the side controlled by the emotive right brain) is more expressive, and surveys in European and American art museums have found that something like 56 percent of men and 68 percent of women in portraits face the left side of the canvas and thereby show more of the left side of the face. Crucifixion scenes of Jesus suffering on the cross showed an even stronger bias, with over 90 percent facing left. (By chance alone, you'd expect closer to 33 percent, since subjects could face left, right, or straight ahead.) And this bias held no matter whether the artists themselves were left- or right-handed. Whether this happens because the sitters prefer to display their more expressive left side or because the artists themselves find that side more interesting isn't clear. But the bias seems universal: it shows up even in high school yearbook photos. A leftward pose also allows the artist to center the sitter's left eye on the canvas. In this position most of her face appears on the canvas's left side, where the face-hungry right hemisphere can study it.

There are exceptions to this leftward bias in portraiture, but even these are telling. The highly ambidextrous Leonardo often broke convention and drew right-facing profiles. But perhaps his most classic piece, the *Mona Lisa,* faces left. Another exception is that self-portraits often face right. Artists tend to paint self-portraits in a mirror, however, which makes the left half of the face *appear* on the right side of the canvas. So this "exception" might actually confirm the bias. Finally, one study found that prominent scientists, at least in their official portraits for the Royal Society in England, usually face right. Perhaps they simply preferred to seem cooler and less emotional—more the stereotypical rationalist.

In contrast to portraits, art in general doesn't show a leftward bias, not in all cultures. In Western paintings, the so-called glance curve—the line the eye naturally follows—does often travel left to right. In art from east Asia, the glance curve more often runs right to left, more in line with reading habits there. A similar bias exists in theater: in Western theaters, as soon as the curtain rises, audiences look left in anticipation; in Chinese theaters, audiences swivel right.

The reason we show a left-right preference for some things (portraits) but not others (landscapes) probably traces back to our evolutionary heritage as animals. Animals can safely ignore most left-right differences in the environment: a scene and its mirror image are more or less identical with regard to food, sex, and shelter. Even smart and discriminating animals—such as rats, who can distinguish squares from rectangles pretty easily—struggle in telling mirror images apart. And human beings, being more animal than not, can be similarly oblivious about left/right differences, even with our own bodies. Russian drill sergeants in the 1800s got so fed up with illiterate peasants not knowing left from right that they'd tie straw to one leg of recruits, hay to the other, then bark, "Straw, hay, straw, hay!" to get them to march in step. Even brainiacs like Sigmund Freud and Richard Feynman admitted to having trouble telling right and left apart. (As a mnemonic, Freud made a quick writing motion with his right hand; Feynman peeked at a mole on his left.) There's also a famous (right-facing) portrait of Goethe showing him with two left feet, and Picasso apparently shrugged at (mis)printed reversals of his works, even when his signature ran the wrong way.

So why then do humans notice any left-right differences? In part because of faces. We're social creatures, and because of our lateralized brains, a right half-grin doesn't quite come off the same as a left half-grin. But the real answer lies in reading and writing. Preliterate children often reverse asymmetric letters like S and N because their brains can't tell the difference. Illiterate artisans who made woodblocks for books in medieval times were bedeviled by the same problem, and their Ƨs and Иs added a clownish levity to dry Latin manuscripts. Only the continual practice we get when reading and writing allows us to remember that these letters slant the way they do. In fact, in all likelihood only the advent of written scripts a few millennia ago forced the human mind to pay much attention to left versus right. It's one more way that literacy changed our brains.

~

Of the three great "proving otherwise"s in Sperry's career, the split-brain work was the most fruitful and the most fascinating. It made Sperry a scientific celebrity and brought colleagues from around the world to his lab. (Although not a schmoozer, Sperry did learn to host a decent party, with folk dancing and a drink called split-brain punch—presumably so named because a few glasses would cleave your mind in two.) The split-brain work entered the popular consciousness as well. Writer Philip K. Dick drew on split-brain research for plot ideas, and the entire educational theory of "left-brain people" versus "right-brain people" derives (however loosely) from Sperry and crew.

Sperry's early proving otherwises probably deserved their own Nobel Prizes, but the split-brain work finally catapulted him to the award in 1981. He shared it with David Hubel and Torsten Wiesel, who'd proved how vision neurons work, thanks to a crooked slide. As lab rats, none of the three had much use for formal attire, and Hubel later recalled hearing a knock on his hotel room door just before the Nobel ceremony in Stockholm. Sperry's son was standing there, his dad's white bow tie limp in his hand: "Does anyone have any idea what to do with this?" Hubel's youngest son, Paul, nodded. Paul had played trumpet in a youth symphony back home and knew all too well about tuxedos. He ended up looping and knotting the bow ties for the geniuses.

Winning a Nobel didn't quench Sperry's ambitions. By the time he won the prize, in fact, he'd all but abandoned his split-brain research to pursue that eternal MacGuffin of neuroscience, the mind-body problem. Like many before him, Sperry didn't believe that you could reduce the mind to mere chirps of neurons. But neither did he believe in dualism, the notion that the mind can exist independently of the brain. Sperry argued instead that the conscious mind was an "emergent property" of neurons.

An example of an emergent property is wetness. Even if you knew every last quantum factoid about H_2O molecules, you'd never be able to predict that sticking your hand into a bucket of water feels wet. Massive numbers of particles must work together for that quality to emerge. The same goes for gravity, another property that surfaces almost magically on macro scales. Sperry argued that our minds emerge in an analogous way: that it takes large numbers of neurons, acting in coordinated ways, to stir a conscious mind to life.

Most scientists agree with Sperry up to this point. More controversially, Sperry argued that the mind, although immaterial, could influence the physical workings of the brain. In other words, pure thoughts somehow had the power to bend back and alter the molecular behavior of the very neurons that gave rise to them. Somehow, mind and brain reciprocally influence each other. It's a bracing idea—and, if true, might explain the nature of consciousness and even provide an opening for free will. But that's a doozy of a *somehow,* and Sperry never conjured up any plausible mechanism for it.

Sperry died in 1994 thinking his work on consciousness and the mind would be his legacy. Colleagues begged to differ, and some of them think back on Sperry's final years (as with Wilder Penfield's late work) with a mixture of disbelief and embarrassment. As one scientist commented, work on the fuzzier aspects of consciousness repels everyone but "fools and Nobel laureates." Nevertheless, Sperry was right about one thing: explaining how human consciousness emerges from the brain has always been—and remains today—*the* defining problem of neuroscience.

CHAPTER TWELVE

The Man, the Myth, the Legend

The ultimate goal of neuroscience is to understand consciousness. It's the most complicated, most sophisticated, most important process in the human brain—and one of the easiest to misunderstand.

September 13, 1848, proved a lovely fall day, bright and clear with a little New England bite. Around 4:30 p.m., when the mind might start wandering, a railroad foreman named Phineas Gage filled a drill hole with gunpowder and turned his head to check on his men. Victims in the annals of medicine almost always go by initials or pseudonyms. Not Gage: his is the most famous name in neuroscience. How ironic, then, that we know so little else about the man.

The Rutland and Burlington Railroad Company was clearing away some rock outcroppings near Cavendish, in central Vermont, that fall, and had hired a gang of Irishmen to blast their way through. While good workers, the men also loved brawling and boozing and shooting guns, and needed kindergarten-level supervision. That's where the twenty-five-year-old Gage came in: the Irishmen respected his toughness, business sense, and people skills, and they loved working for him. Before September 13, in fact, the railroad considered Gage the best foreman in its ranks.

As foreman, Gage had to determine where to place the drill holes, a job that was half geology, half geometry. The holes reached a few feet deep into the black rock and had to run along natural joints and rifts to help blow the rock apart. After the hole was drilled, the foreman sprinkled in gunpowder, then tamped the powder down, gently, with an iron rod. This completed, he snaked a fuse into the hole. Finally an assistant poured in sand or clay, which got tamped down hard, to confine the bang to a tiny space. Most foremen used a crowbar for tamping, but Gage had commissioned his own rod from a blacksmith. Instead of a crowbar's elongated S, Gage's rod was straight and sleek, like a javelin. It weighed 13¼ pounds and stretched three feet seven inches long (Gage stood five-six). At its widest the rod had a diameter of 1¼ inches, although the last foot—the part Gage held near his head when tamping—tapered to a point.

Around 4:30 Gage's crew apparently distracted him; they were loading some busted rock onto a cart, and it was near quitting time, so perhaps they were a-whooping and a-hollering. Gage had just finished pouring some powder into a hole, and turned his head. Accounts differ about what happened next. Some say Gage tried to tamp the gunpowder down with his head still turned, and scraped his iron against the side of the hole, creating a spark. Some say Gage's assistant (perhaps also distracted) failed to pour the sand into the hole, and when Gage turned back he smashed the rod down hard, thinking he was packing inert material. Regardless, a spark shot out somewhere in the dark cavity, and the tamping iron reversed thrusters.

Gage was likely speaking at that instant, with his jaw open. The iron entered point first, striking Gage point-blank below the left cheekbone. The rod destroyed an upper molar, pierced the left eye socket, and passed behind the eye into his brainpan. At this point things get murky. The size and position of the brain within the skull, as well as the size and position of individual features within the brain itself, vary from person to person—brains vary as much as faces do. So no one knows exactly what got damaged inside Gage's brain (a point worth remembering). But the iron did enter the underbelly of his left frontal lobe and plow through the top of his skull, exiting where babies have their soft spots. After parabola-ing upward—it reportedly whistled as it flew—the rod landed twenty-five yards distant and stuck upright in the dirt, mumblety-peg-style. Witnesses described it as streaked with red and greasy to the touch from fatty brain tissue.

The rod's momentum threw Gage backward and he landed hard. Amazingly, though, he claimed he never lost consciousness, not even for an eyeblink. He merely twitched a few times on the ground, and was talking again within a few minutes. He walked to a nearby oxcart and climbed in, and someone grabbed the reins and giddyupped. Despite the injury Gage sat upright for the mile-long trip into Cavendish, then dismounted with minimal assistance at the hotel where he was lodging. He took a load off in a chair on the porch and even chatted with passersby, who could see a funnel of upturned bone jutting out of his scalp.

Two doctors eventually arrived. Gage greeted the first by angling his head and deadpanning, "Here's business enough for you." Doctor one's "treatment" of Gage hardly merited the term: "the parts of the brain that looked good for something, I put back in," he later recalled, and threw the "no good" parts out. Beyond that, he spent much of his hour with Gage questioning the veracity of the witnesses. *You're sure? The rod passed* through *his skull?* On this point the doctor also queried Gage himself, who—despite all expectation—had remained utterly calm and lucid since the accident, betraying no discomfort, no pain, no stress or worry. Gage answered the doctor by pointing to his left cheek, which was smeared with rust and black powder. A two-inch flap there led straight into his brain.

Finally, Dr. John Harlow arrived around 6 p.m. Just twenty-nine years old, and a self-described "obscure country physician," Harlow spent his days treating people who'd fallen from horses and gotten in carriage accidents, not neurological cases. He'd heard nothing of the new theories of localization simmering in Europe and had no inkling that, decades later, his new patient would become central to the field.

Like everyone else, Harlow didn't believe Gage at first. *Surely, the rod didn't pass* through *your skull?* But after receiving assurance that it had, Harlow watched Gage lumber upstairs to his hotel room and lie down on the bed—which pretty much ruined the linens, since his upper body was one big bloody mess. As for what happened next, readers with queasy stomachs should skip to the next paragraph. (I'm not kidding.) Harlow shaved Gage's scalp and peeled off the dried blood and gelatinous brains. He then extracted skull fragments from the wound by sticking his fingers in from both ends, Chinese-finger-trap-style. Throughout all this, Gage was retching every twenty minutes, mostly because blood and greasy bits of brain kept slipping down the back of his throat and gagging him. The violence of the heaving also caused "half a teacupful" of brain to ooze out the exit wound on top. Incredibly, even after tasting his own brains, Gage never got ruffled. He remained conscious and rational throughout. The only false note was Gage's boast that he'd be back blasting rocks within two days.

The bleeding stopped around 11 p.m. Gage's left eyeball was still protruding a good half inch, and his head and arms remained heavily bandaged (he had flash burns up to his elbows). Harlow nevertheless allowed visitors the next morning, and Gage recognized his mother and uncle, a good sign. He remained stable over the next few days thanks to Harlow's diligent care, which included fresh dressings and cold compresses. But just when Harlow grew hopeful that Gage would survive, his condition deteriorated. His face puffed up, his brain swelled, and the wound, no doubt due to something beneath Harlow's fingernails, developed a mushrooming fungal infection. Worse, as his brain continued swelling, Gage started raving, demanding that someone find his pants so he could go outside. He soon lapsed into a coma, and at one point a local cabinetmaker measured him for a coffin.

Gage would indeed have died—of intracranial pressure, like Henri II three centuries before—if Harlow hadn't performed emergency surgery and punctured the tissue inside his nose to drain the wound of pus and blood. Things were touch and go for a few weeks afterward, and Gage did lose sight in his left eye. (The lid remained sewn shut the rest of his life.) But he eventually stabilized and returned home to Lebanon, New Hampshire, in late November. In his case notes Harlow downplayed his role here and even quoted Ambroise Paré: "I dressed him, God healed him." In reality it was Harlow's dedicated care and his bravery in performing an emergency operation—something Paré had refused to do with Henri—that saved Phineas Gage.

Or did it? Harlow kept Gage alive, but Gage's friends and family swore that the man who came home to Lebanon was not the same man who'd left Lebanon months before. True, most things were the

Daguerreotype of Phineas Gage. (Collection of Jack and Beverly Wilgus)

same. He suffered some memory lapses (probably inevitable), but otherwise his basic mental faculties remained intact. It was his personality that had changed, and not for the better. Although resolute in his plans before the accident, this Gage was capricious, almost ADD, and no sooner made a plan than dropped it for another scheme. Although deferential to people's wishes before, this Gage chafed at any restraint on his desires. Although a canny businessman before, this Gage lacked money sense: Harlow once tested Gage by offering him $1,000 for some random pebbles that Gage had picked out of a riverbed; Gage refused. And although a courteous and reverent man before, this Gage was foul-mouthed. (To be fair, you'd probably swear, too, if an iron rod had rocketed through your skull.) Harlow summed up Gage's personality changes by saying, "The equilibrium or balance... between his intellectual faculties and his animal propensities seems to have been destroyed." More pithily, friends said that Gage "was no longer Gage."

Despite his stellar work record, railroad managers refused to reinstate Gage as foreman. So he began working odd jobs on farms, and even exhibited himself and his tamping iron—his constant companion now—for spare cash in P. T. Barnum's museum in New York, staring back at the audience with his one good eye. (For an extra dime skeptics could part his hair and gape at the one-inch-by-two-inch soft spot in his skull, beneath which his brain still pulsated.) After leaving Barnum's employ he indulged a newfound love of horses and became a stablehand and coach driver in New Hampshire. He also felt drawn to children, and on visits home would spin wild—and wholly untrue—yarns for his nieces and nephews about his supposed adventures. Whether this was a simple love of tall tales or, consistent with his frontal lobe damage, a sign of confabulation, no one knows.

Ironically, Gage's own life story soon became something of a tall tale. Not immediately: Gage lived a mostly anonymous life after his accident. But in the decades after his death, rumors began to circulate about him—some of them plausible, some of them pretty warped, all of them probably false. One claimed that Gage developed a drinking problem and started scrapping and brawling in taverns. Another claimed that he turned into a scam artist: he supposedly sold the exclusive, posthumous rights to his skull to a certain medical school, then sold the same rights to another school, and another, and another, skipping town and pocketing the cash each time. One source even had Gage living for a dozen years with the tamping iron still impaled in his noggin.

More important for neuroscience, there's a dearth of hard detail about the personality changes he experienced. We simply don't know how Gage spent most of his postaccident life, nor what his behavior was really like. Harlow's case reports do make it clear that Gage changed somehow; but Harlow focuses more on his blue language and irrational attachment to pebbles than on the things neuroscientists would investigate today, like Gage's foresight, emotional capacity, or ability to complete a sequence of steps. As a result of all this, Gage's life has become as much legend as fact, and the most tantalizing questions about him—how did his mind work now? did he see himself differently? did he recover any lost skills?—remain unanswered.

Nevertheless, all is not lost. If we're careful, there are some recent cases in neuroscience that can at least get at those questions. There are "modern Phineas Gages" who can help us glimpse how, when the iron rod finished remodeling Gage's brain, his mind might have changed in response.

~

Of all the incredible details about Gage's accident, perhaps the most incredible is his claim that he never lost consciousness. Still, in light of modern research, the claim makes some sense.

Neuroscientists of yesteryear scoured every last cranny inside the brain for the seat of human consciousness. Modern neuroscientists search for something different. As one put it, "Consciousness isn't a thing in a place; it's a process in a population."* That is, consciousness isn't localized: it emerges only when multiple parts of the brain hum in harmony.

Some of these parts provide basic infrastructural support. A web of neurons in the brainstem called the reticular formation controls sleeping and waking cycles, and acts like the power switch for consciousness. If it suffers damage, basic bodily processes like breathing and digestion continue, but the brain can't "boot up" its higher faculties. Lesser injuries, like concussions, can also send ripples through the brain that disrupt the reticular formation and cause blackouts. Gage's injury, by contrast, was focal: however gruesome, the damage was confined to a small tunnel of tissue, without a devastating shock wave of trauma. As a result, his reticular formation escaped unscathed, and his consciousness might never have experienced any hiccups.

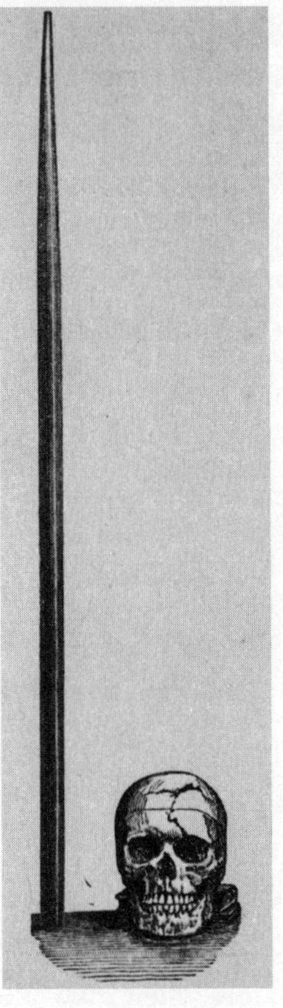

Drawing comparing the size of Gage's skull and tamping iron. (National Library of Medicine)

However important for supporting consciousness, though, the reticular formation and related structures don't actually stir consciousness to life. That responsibility falls more to the thalamus and prefrontal parietal network.

The thalamus, at the core of the brain, brokers information. It receives information from all over the brain, analyzes it, then relays it around—patching different parts of the brain together like an old-time telephone operator. And for whatever reason, damage to the thalamic relay centers can obliterate consciousness, leading to what's called a vegetative state. Unlike coma victims, vegetables remain awake, but they cannot focus on anything or engage in any higher thinking. Their minds drift listlessly from moment to moment, leaves in an indifferent wind. You can also become a vegetable if you suffer damage to the prefrontal parietal network, which (truth in advertising) consists of a patch of frontal cortex, a patch of parietal cortex, and the connections between them. These two patches almost always light up in tandem when we pay close attention to something, an important aspect of consciousness. In sum, the thalamus and prefrontal parietal network don't ignite consciousness all by themselves, but they do keep the fire stoked.

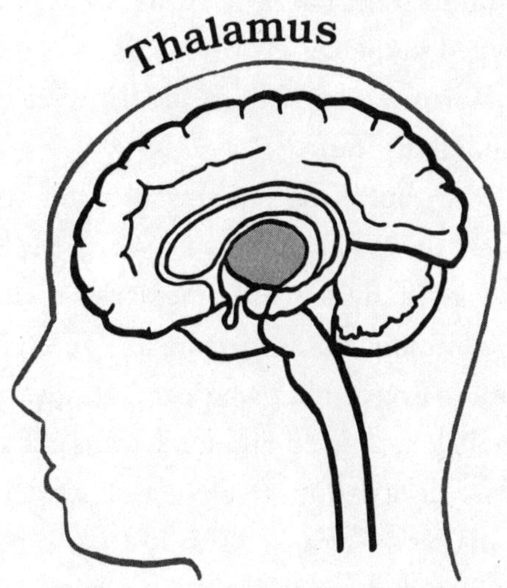

Another prerequisite for consciousness is short-term memory, since consciousness requires you to keep track of things from minute to minute. Most amnesiacs, like H.M. and K.C., do have a working short-term memory and a normal moment-to-moment consciousness. But there are people out there with even more severe amnesia, like English musician Clive Wearing, whose consciousness functions differently.

Wearing made a name for himself in the 1970s as a classical musician and conductor; his concerts of Renaissance music—which re-created everything from the costumes the musicians wore to the meals they ate before performances—have been described as "the next best thing to going back in time." He also arranged the score for a BBC radio broadcast celebrating Charles and Diana's wedding in 1981. Clive himself got hitched two years later, but in March 1985, at age forty-six, he came down with a prolonged "flu" and headache; doctors diagnosed meningitis, which had been going around London that week. He became lethargic and irritable, and at one point wandered outside, got lost, hailed a cab, and couldn't remember his address. The driver dumped him at a local police station, where his wife eventually found him. He suffered on for six more days before finally being dragged to the hospital. Doctors there diagnosed our old friend the herpes virus, and he began suffering seizures and lapsing into and out of consciousness.

Wearing pulled through and remains alive today. But he suffered heavy limbic-system damage and woke up with no episodic (personal) memory whatsoever. Many semantic memories disappeared, too: he couldn't define common words like "tree," "eyelid," or (fittingly) "amnesia"; he couldn't remember who penned *Romeo and Juliet;* and he once ate a whole lemon, rind and all, because he didn't recognize what it was. Most devastating of all—and unlike virtually every other known amnesiac—Wearing also lost his short-term working memory. When he turned his head, people's shirts seemed to change color; when he blinked, the cards in his solitaire game rearranged themselves. Especially at first, his memories lasted no longer than his sensory perceptions did.

As a result, Wearing lost all sense of continuity between past and present: as far as he knew, no other day had ever existed. And however strange it sounds, he interpreted this break with the past as evidence that he'd just "woken up." That is, he started claiming, incessantly, every few minutes, with the zeal of an evangelist, that he'd just become conscious for the very first time. To be clear: Wearing wasn't actually blacking out or anything; anyone watching him would have seen that he remained awake moment to moment. But in his own mind, based on the little evidence available to him, he could only conclude that the past few seconds were his first-ever moments of consciousness. This ecstatic rebirth reoccurred dozens of time every day.

This obsession with consciousness comes through most clearly in his diaries. He started keeping a diary in 1985, to provide an anchor for his past—proof he even had a past. Instead, Wearing filled whole pages with entries like:

> ~~*8:31 AM: Now I am really, completely awake.*~~
> ~~*9:06 AM: Now I am perfectly, overwhelmingly awake.*~~
> *9:34 AM: Now I am superlatively, actually awake.*

And so on. Every few minutes, the rapture of having just become conscious would overwhelm him and compel him to record the moment. (A few times, when he couldn't find the diary right away, he grabbed a pen and recorded his epiphany on the walls or furniture.) But because he'd only just woken up—just now—the old entries were clearly false. Hence, he struck them out.

Wearing has dozens of diaries littered with such entries, each one denying, with incredible adverbial dexterity, that he'd ever been awake before. And as you might suspect, Occam's razor cannot kill this delusion: he can even recognize his handwriting in the struck-out passages, but any suggestion that he therefore probably wrote them can send him into a rage. Old videos of him playing the piano do the same. He once again recognizes himself in them, but denies he was actually conscious at the time. When asked the obvious follow-up question—what, then, was going on in your head during those videos?—he might erupt, *How the hell should I know? I've only just woken up.*

So why does Wearing lose consciousness over and over, while Gage never lost it at all? Again, we know the rough answer with regard to Gage: the skinny tamping iron must have skirted all the regions that help produce consciousness, or it would have been lights out. And lest you dismiss the claim that Gage stayed awake as nineteenth-century credulity, there are modern reports of people getting impaled with metal rods or shafts and remaining cognizant,* too. Gage was nothing special here.

Wearing's case is tougher to understand. His consciousness circuits certainly work to some extent, since he can recognize that he's conscious at any one moment. But part of being conscious is *maintaining* that awareness over time, and whatever structures in the brain support that function seem to be draining every few seconds, like a battery that can't hold a charge. So while Wearing never quite sinks into a vegetative state, he never quite emerges into full, sustained consciousness, either. This might make sense if Wearing's thalamus, prefrontal parietal network, or reticular formation had suffered damage, but they actually look okay on brain scans. So scientists are reduced to guessing. Perhaps some region that connects those structures together suffered damage. Perhaps those structures suffered damage that brain scans can't pick up. (Wearing does confabulate, a sign of frontal lobe impairment, and some neuroscientists have pegged his endless babbling and "incontinent punning" as another frontal lobe disorder, *Witzelsucht*, literally, the joking disease.) Perhaps the damage to individual structures matters less than the overall, brain-wide extent of his damage. Or perhaps Wearing's troubles can be traced to something we don't yet understand, something that plays an unsuspected role in consciousness. Nor do we understand why other amnesiacs escaped his fate. H.M. and others do sense the present constantly slipping away from them, seeping into indistinctness, and it unnerves them. But unlike Wearing, they don't deny that their pasts exist. Only Wearing loses the continuity and continually "pops awake."

In the end, Gage and Wearing sit on a spectrum, the tenacity of Gage's consciousness on one end, the fragility of Wearing's on the other. You certainly couldn't call Gage lucky, but his focal damage at least spared his consciousness. Wearing, meanwhile, enjoys neither the gift of full mental awareness nor the release of permanent oblivion. Instead, his own brain torments him with an almost mythological malice. Like Sisyphus's boulder, as soon as he gets a purchase on his consciousness, it slips away. Like Prometheus's liver, it grows back every few seconds, only to be torn out again.*

~

The comment by his loved ones that Gage "was no longer Gage" after the accident brings up another point worth unpacking. To friends and family, Gage had changed, clearly. But how did Gage himself understand these changes? Was his sense of self transformed or diminished? Sadly, Gage didn't record his thoughts on this (or any) subject. But again, we can infer some things about his sense of self from other cases of brain damage.

The annals of neuroscience contain some pretty distorted views of the self. Victims of Cotard syndrome are convinced they're corpses. Other deluded folk swear they have three arms or legs. H.M. never matured beyond age seventeen in his own mind. (When handed a mirror, he would stare nonplussed at his wrinkles and gray hair and deadpan, "I'm not a boy.") Other amnesiacs forget things you wouldn't think possible, even basic biological functions. Aleksandr Luria, the Russian neuroscientist who studied the memory freak Shereshevsky, wrote another "neurological novel" about a soldier named Zazetsky, who took a bullet to the parietal lobe while fighting off Nazis near Belarus in 1943. The parietal lobe helps monitor bodily sensations, and when his parietal lobe got shredded, Zazetsky forgot how to go to the bathroom. He would feel a bulge in his sphincters and know something was up, but couldn't recall what to do next.

Still, even the most desperate amnesiacs never forget themselves—never forget, deep down, who they are. For instance, most amnesiacs can describe their own personalities: they know they're generous or impatient or whatever, even if they can't recall a single time they displayed that trait. They can also tap into their core selves by drawing on different types of memory. Clive Wearing can still sight-read music and play the piano, since those skills draw on his procedural (unconscious) memories. And for whatever reason, being a musician is so deeply rooted within him that those procedural memories can resurrect something of his old, lost self: as soon as he strikes the first chord, the momentum of the phrases keeps him intact, dragging him along and providing a coherence and unity he otherwise lacks. It's as if he slipped through a wormhole into an alternative dimension where his brain circuits never suffered damage. After the final note, of course, he's ejected from that world. And the bewilderment and disappointment of finding himself lost again often causes a swell of emotion so intense that his body begins to convulse. But for the entire étude or rondo, Clive is Clive again.

In addition to music, Wearing's emotional memories provide an anchor. He lost his memory within two years of getting married, and in the thirty years since, he hasn't lost an iota of passion for his Deborah. Every time—every single time—she visits him at his nursing home, he erupts in joy. If she steps out to the ladies' room, he might crumble—then erupt again when she returns. And for a few years, as soon as Deborah left for the day, Clive would begin leaving messages on her answering machine, begging to know why she never visited. "Hello, love, 'tis me, Clive. It's five minutes past four... I'm awake for the first time..." *Beep.* "Darling?... It's a quarter past four and I'm awake now for the first time..." *Beep.* "Darling? It's me, Clive, and it's eighteen minutes past four and I'm awake..." However flattering—would that we were all loved so much!—Deborah admitted that she sometimes had trouble faking enthusiasm for yet another "reunion." But there's no denying that Clive is tapping into his inner core here—something that he'll never abandon and that will never abandon him.*

The tenacity of the self is revealed even more clearly in another case of distorted consciousness, that of Tatiana and Krista, conjoined twins born in British Columbia in 2006. Surgeons declined to separate the girls at birth because they have, essentially, a Siamese brain, with their skulls fused together. (The girls face the same direction, with Tatiana on their right. They cannot see each other but can walk just fine, leaning toward each other like two legs of a triangle.) Inside their joint skull, a stretch of axon cables connects their thalami. As far as doctors know, this "thalamic bridge" is unique in medical history, and as Tatiana and Krista get older and more articulate, they're showing some amazing behaviors. They often speak simultaneously, like two stereo speakers, and each can taste what's in the other's mouth. Prick one for a blood test, and the other winces. Put them down to bed, and they fall asleep in unison, and possibly dream together.

In other words, each girl has access to the other's consciousness, and neither girl distinguishes clearly between her own thoughts and sensations and those of her sister. Their use of pronouns reflects this ambiguity. They use "I" in odd situations: hand each of them a single piece of paper, for instance, and they'll announce, "I have two pieces of paper." And they never say "we," either, as if their single thalamus unites them. The girls do have other brain anomalies: each has a tiny corpus callosum, and Tatiana's left hemisphere and Krista's right hemisphere (i.e., the hemispheres between them) never developed fully. But the thalamic bridge is probably what produces their hybrid consciousness.

And yet, despite sharing in each other's consciousness, each girl shows strong signs of individuality. With regard to food, Krista breaks out in hives whenever she eats canned corn; Tatiana doesn't. And while Krista likes catsup, Tatiana hates it, and tries to scrape it off her tongue whenever Krista eats it. The girls also fight as if two separate people—punching each other, gouging eyes, pulling hair. This can lead to some *Three Stooges* absurdities, like when one girl slaps the other, then grabs her own face in pain. But they apparently feel distinct enough to attack one another as foreign. One twin sometimes even says, out of the blue, as if affirming it, "I am just me." Of course, her sister often undermines her a moment later by echoing, "*I* am just *me*." (Shades of *The Shining* twins there.) But there's clearly a need, an instinct, to declare their independence.

Psychologists of a certain bent have always denied that people have a fixed core, a fixed self. And given how much we change roles and shape-shift mentally from one situation to another, depending on the social milieu and whom we're talking to, these psychologists have a point. Neurologically, though, we do seem to have a core brain circuit that defines and establishes a self. This sense of self weaves together many different strands: autobiographical memories; physical looks; a sense of continuity through time; a sense of personal agency; knowledge of our own personality traits; and so on. But like a tapestry, the self doesn't depend on the integrity of any one strand alone: K.C. lost his autobiography, the *mutilés* of World War I lost their faces, Clive Wearing lost all continuity, alien hand victims lost personal agency. And yet all of them retained a sense of self. Like consciousness, the self is less a thing in a place than a process in a population—and that makes the self tenacious, stronger than any of the vicissitudes of life.

So in all likelihood, if you'd asked him, Phineas Gage would have told you that he still felt like Phineas Gage. Always had.

~

The most important details of Gage's case involve the psychological changes he underwent, because of damage to the front of his frontal lobes. Unfortunately, this is also the area where hard facts are hardest to come by. No one ever performed any psychological evaluations of Gage, and beyond saying "the prefrontal area," we don't even know what regions of his brain suffered damage, either from the tamping iron or from the subsequent swelling and infection. Modern neuroscientists have nevertheless found it irresistible to read between the lines of what Gage's doctors strictly reported and to equate Gage with modern patients.

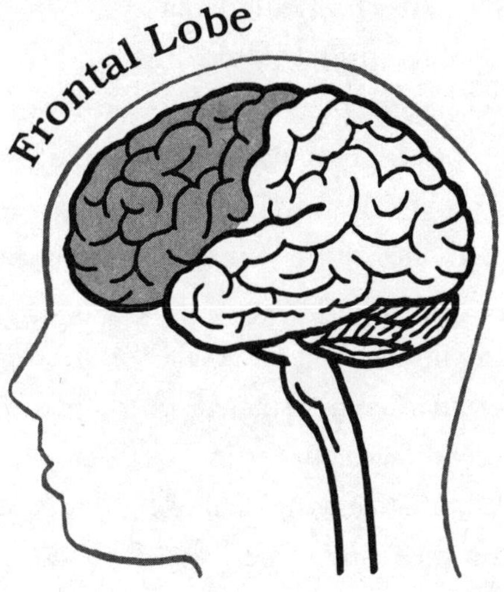

The patient most commonly called a "modern Phineas Gage" is Elliot, whom we met in the emotions chapter. (After a tumor crushed his frontal lobes, Elliot spent hours deciding what restaurant to eat at or how to sort tax documents. He also lost his nest egg in a dubious investment.) Neuroscientists associate Elliot and Gage because they both displayed probably the classic symptom of prefrontal lobe damage—personality changes. People who suffer prefrontal damage rarely die from it, and their senses, reflexes, language, memory, and reasoning survive intact. Indeed, a stranger who chatted with Gage or Elliot for a minute probably wouldn't have noticed anything amiss. But anyone who knew and cared about them could spot the differences immediately: the mental changes were as obvious as a facial scar. Prefrontal damage might not kill people, but it can kill what we cherish most about them.

Beyond their altered personalities, however, it's hard to know how closely Gage's and Elliot's stories actually parallel each other. On the one hand, the resemblance seems tantalizing, more than enough for, say, a good lawyer to convince you of the similarities: Neither man could resume his job after his brain damage, and both betrayed a sudden lack of money sense, Elliot in making bad investments, Gage in refusing to part with some pebbles for $1,000. Both showed a lack of embarrassment in social situations: Gage reportedly swore like a pirate and let people dig through his hair for dimes; Elliot confessed every last squalid detail about his life without a hint of shame, right down to his moving back in with his parents in his forties. Both showed an attachment to inanimate objects: Gage carried his tamping iron everywhere; Elliot hoarded newspapers, dead houseplants, and empty cans of frozen orange-juice concentrate. Both men seemed enslaved to their impulses: Elliot's marrying a hooker sounds an awful lot like Gage's doctor's comment about the "animal passions" winning out in his patient. Both men hurt loved ones with their callousness, and both showed possible signs of emotional disturbance: Elliot's feelings flatlined, and nothing—not music, not painting, not even politics he despised—could rouse him; Gage, in the aftermath of his accident, remained unflappable, eerily indifferent, as if (as some modern commentators claim) he'd been lobotomized.

All that said, you can also read Gage's story another way, in which case the Elliot comparisons seem exaggerated and unfair. We actually know very little about Gage's mental life overall, and what we do know seems ambiguous, even cryptic, if we read carefully. Take the comment about Gage's sudden "animal passions." Sounds impressive, but what does that mean? Did he eat or sleep too much? Demand sex? Howl at the moon? It depends entirely on interpretation. As for his supposed attachment to objects, Gage schlepped his tamping iron around everywhere, sure, but can you blame him? Attachment to the rod that remodeled your brain is surely more rational than hoarding cans of frozen orange juice. As for Gage's emotions, beyond his indifference right after the accident—which could be due to shock—we know nothing, zero, about his emotional life in later years. And while Gage did have trouble sticking to plans and seemed to lose the impulse control that prevents nice people from cursing in public, a saucy "hell" or "damn" in conversation hardly makes him an antebellum Elliot.

Indeed, some modern historians* have argued, forcefully, that while Gage did show signs of frontal lobe damage right after his accident, he also—unlike Elliot—seems to have recovered some of his faculties over the next decade. He never became the Phineas Gage of old (no hope of that), but some of his negative traits either diminished or disappeared, possibly because his brain proved plastic enough to recover lost functions.

After his stints in Barnum's museum and a New Hampshire horse stable, Gage took off for Chile in 1852, probably following a gold rush. He was seasick the whole voyage. Once ashore, he found work driving a carriage and shuttling passengers along the rugged, mountainous trails between Valparaiso and Santiago. Considering his brain damage, his success in this job—he did it for seven years—beggars belief. He likely drove a team of six horses, which demanded no little dexterity, since he had to control each horse separately. Rounding a bend without tipping the coach over, for instance, required slowing down the inner three horses a touch more than the outer three, simply by tugging on their reins with varying amounts of pressure. (Imagine driving a car while steering all four wheels independently.) Furthermore, the trails were crowded, forcing him to make quick stops and dodges, and because he probably drove at night sometimes, he would have had to memorize their twists and turns and fatal drop-offs, all the while keeping an eye out for banditos. He also likely cared for his horses and (contra the claim that he lacked all money sense) collected passenger fares. Not to mention that he presumably picked up a soupçon of español in Chile. You wonder how many of Gage's passengers would have climbed in had they known about their one-eyed driver's little accident a few years before, but he seems to have handled himself fine, far better than Elliot ever did.

That Gage carved out a life for himself in Chile doesn't mean that his brain recovered fully. It merely suggests that his brain recovered somewhat. As we've seen, the brain's neural circuits can rewire themselves in certain circumstances, and perhaps Gage retained enough of his frontal lobes (especially on the right side) to compensate for his lost social and executive skills. At the very least, Gage didn't deteriorate into the drunken sociopath that many modern accounts make him out to be.

One factor that may have helped Gage thrive (and that may explain why Elliot didn't) was the routinized nature of Gage's work. He likely arose each day before dawn to prep the horses and carriage, then spent the next thirteen hours driving the same road from Valparaiso to Santiago and back. As noted, victims of frontal lobe damage often have trouble finishing tasks, especially open-ended tasks, because they get distracted or overwhelmed. But all Gage had to do was keep driving forward until it was time to turn around, and every day unfolded pretty much the same way. This introduced structure into his life and probably helped him avoid a life of dissolution. Gage might not have been Gage anymore, but he wasn't a wastrel.

Still, he couldn't outrun his brain damage entirely—and when it did catch up to him, the end was swift. Increasingly poor health forced him to quit Chile, and in 1859 he caught a steamer to San Francisco, near where his family had moved. After a few months of rest, he found work as a farm laborer and seemed to be doing better, until a punishing day of plowing in early 1860 wiped him out. He had a seizure the next night over dinner. More followed.

Gage tried gamely to continue working during this spell, but he suddenly became restless and capricious, and began drifting from farm to farm, always finding a reason to quit his current job. Finally, at 5 a.m. on May 20, while resting at his mother's home, he had a fit more violent than any before. In fact the fits never really stopped after that, and Gage entered a state called status epilepticus—a continuous seizure. He died on May 21, age thirty-six, having survived his accident by almost a dozen years. His family buried him two days later, presumably with his beloved tamping iron. To the inestimable loss of the world, there was no San Francisco Broca to preserve his brain.

Gage's story might have ended there—an obscure small-town tragedy, little more—if not for Dr. John Harlow. Harlow had lost track of Gage after Gage shipped off to Chile in 1852. (Amid other distractions, Harlow got caught up in politics, and he later won a seat in the Massachusetts state senate.) Nevertheless, Gage's story kept nagging Harlow; he couldn't shake the thought that his former patient had more to teach the medical world. So when Harlow learned the address of Gage's mother in 1866 (through some unspecified "good fortune"), he wrote to California straightaway for news.

Although panged that they'd not arranged for an autopsy, Harlow exchanged some letters with the Gages and milked them for details of Gage's life. He then prevailed upon Gage's sister, Phebe, to have the grave opened in late 1867 to salvage Gage's skull. The exhumation sounded like quite a to-do, with Phebe, her husband, their family doctor, the city mortician, and even San Francisco's mayor, one Dr. Coon, all present to peek inside the coffin. Gage's family hand-delivered the skull and tamping iron to Harlow in New York a few months later. After interviewing the family and studying the skull, Harlow wrote a detailed case report on Gage in 1868, including most of what we know about his psychological transformation. His work on Gage complete, Harlow donated the skull and tamping iron to an anatomical museum at Harvard University, where they remain today.

Harlow persisted in tracking Gage down and documenting his story in part because he assumed that posterity would forget about Gage otherwise. But in the two decades following Gage's accident, neuroscience had changed considerably. Europe was suddenly frothing over with debates about brain localization, and although most Europeans didn't take American science seriously, the singularity of Gage's injuries—*You are sure, Yan-kee? Ze rod went* zhrough *his skull?*—proved too fascinating to ignore. Over the next few decades neuroscientists began debating Gage's case in earnest.

In reality, the dearth of hard details about Gage probably secured his fame, since it left infinite room for interpretation and bickering. Gage became—and remains today—something of a Rorschach blot for neuroscientists, an indication of the passions and obsessions of each passing era. Phrenologists explained some of Gage's symptoms, like his cursing, by noting that his "organ of veneration" had been blown to bits. Roberts Bartholow cited Gage in defense of his experiments on the exposed brain of Mary Rafferty, for if Gage could survive having his skull reamed, Bartholow argued, how was he to know that a little electricity would kill? Neurosurgeons, oddly, saw Gage as an inspiration. Whatever did or didn't change inside him, Gage proved that people could at least survive extensive loss of brain tissue. This reassured surgeons during an era of appallingly high death rates, and justified the surgical approach to treating certain brain disorders.

Above all, Gage got dragged into that all-time classic debate in neuroscience—over localization and the supposed seat of our humanity. Many antilocalizationists actually seized upon Gage as evidence of a unified and nonlocalized brain, a counterweight to the likes of Tan and Lelo. They emphasized, first of all, that despite the widespread damage, Gage had retained most of his mental faculties: he could still reason, remember, recognize faces, and learn new skills. Furthermore, due to a misunderstanding, the antis thought that the tamping iron had actually destroyed the rear frontal lobes—the very regions that Broca and other localizationists were passing off as speech and motor centers. Since Gage never lost those skills, the antis argued that localization theory must be bunk.

Localizationists parried. While they conceded that Gage had retained most of his mental faculties, those talents might simply be located in other lobes. What's more, they dug up an experiment from 1849 in which a doctor had drilled a hole in a cadaver's skull to determine the path of the tamping iron through Gage's head. This sounds a bit like Henri II's doctors battering the skulls of decapitated criminals with the lance butt, but this experiment actually yielded useful information: it proved that the rod had almost certainly missed Gage's speech and motor centers, rendering that objection mute. Most important, localizationists noted that—whatever else got spared—Gage's personality *had* changed, drastically. The human mind isn't just memory plus language plus reasoning plus sensory data, all working independently: those modules have to come together and find a common expression. They do so in the frontal lobes, which serve as a hub to integrate those isolated talents. And when that hub was destroyed, Gage lost something essentially human. He was no longer Gage.

The localization arguments eventually carried the day. There was simply too much evidence that the damage to Gage's frontal frontal lobes had remade his personality. And from there it was but a small step to one of the founding doctrines of modern neuroscience: that brain and mind are interlocked. Somewhere inside our gray and white matter, we can indeed find mere flesh that, if sparked a certain way or drenched in a certain soup, can produce generosity, patience, kindness, persistence, common sense—or a lack of any of those things. Gage's case alone didn't drag neuroscience to this conclusion. But after him, scientists had real proof that the glories of the human mind arise directly from the intricacies of the human brain. No matter what parts of his life remain murky or debatable, Gage remains probably the most important case in the history of neuroscience because his story pointed us toward that truth.

~

Gage's story retains its grip on us for other reasons as well. Stories probably mean more to neuroscience than to any other scientific field—and as we've seen throughout this book, they're not always the easiest stories to take. Some are in fact downright difficult to sit through, striking a little too close to home. Unlike with other fields, any one of us could make a vital contribution to neuroscience someday, through no fault of our own. Our names (or at least initials) might be immortalized in textbooks, and like many other facets of neuroscience, that's both amazing to think about and scary.

With Gage, it's fitting that his life has been transformed into legend. He and so many others from the history of neuroscience—the kuru cannibals, the pituitary giants, even blind James Holman—do at times seem like characters from myths or fairy tales. And like fables, their stories have taught us an awful lot. We now know how our neurons fire and exchange neurotransmitters. We know how circuits chirp and whir upon seeing a familiar face. We know what underlies our urges and animal drives, and from those building blocks we can reconstruct how we reason and move and communicate. Above all, we know that there's a physical basis for every psychological attribute we have: if just the right spot gets damaged, we can lose just about anything in our mental repertoire, no matter how sacred. And although we don't fully understand the alchemy that transforms the buzzing of billions of cells into a spritely, creative human mind, new tales continue to pull the curtain back a little farther.

Perhaps even more important than the science, these stories enrich our understanding of the human condition—which is, after all, the point of stories. Whenever we read about people's lives, fictional or non-, we have to put ourselves into the minds of the characters. And honestly, my mind has never had to stretch so far, never had to work so hard, as it did to inhabit the minds of people with brain damage. They're recognizably human in so many ways, and yet still somehow off: Hamlet seems transparent next to H.M.

But that's the power of stories, to reach across that divide. These people's minds don't work quite like ours, it's true. We can still identify with them, though, on a basic, human level: They want the same things we want, and endure the same disappointments. Feel the same joys, and suffer the same bewilderment that life got away from them. Even their tragedies provide some solace, for we know that if any one of us were to suffer a catastrophic injury—or succumb to a common plague of old age, like Alzheimer's or Parkinson's—our minds would cling to our inner selves with the same tenacity. The *you* in you won't disappear.

There are a lot of tales of injury and woe in this book. But there's a hell of a lot of resiliency, too. We're all fragile, and we're all very, very strong. Even that paradigmatic example of a life falling apart because of a brain injury, Phineas Gage, might have recovered more than most scientists ever hoped. No one's brain gets through life unscathed. But the thing about the brain is that, despite what changes, so much remains intact. Despite all the differences between different people's minds, that's one thing we all share. After his accident, friends and family swore that Phineas Gage was no longer Phineas Gage. Well, he was and he wasn't. And he was all of us, too.

Acknowledgments

The book you're holding in your hand was the product of many, many different people's brains, and I feel lucky for having had the chance to tap into their collective consciousness and collect the results. Everyone contributed something important, and if I've left anyone off this list, I remain thankful, if embarrassed.

Once again, a big thank-you to my loved ones. My parents, Gene and Jean, have been there for me literally my whole life, and have taken my occasional writing about them with good humor. (That's why I'm not going to point out that my mother actually failed the happy/sad face test on page 324. She thinks backwards.) The same goes for my siblings, Ben and Becca, two of the best people I know. And I'm happy to add some little'uns, Penny and Harrison Schultz, to the list this time. All my friends in Washington, D.C., and South Dakota and around the country have helped me get through some rough times, and I'm happy to share the good ones with them still.

My agent, Rick Broadhead, loved this idea from the get-go, and helped steer it to a great finish. And I thank, too, my editor John Parsley, whose encouragement and insight teased out the best I had inside me. I spent many hours writing before I knew John, but he taught me what I know about writing a *book*. Also invaluable were others at and around Little, Brown who've worked with me on this book and others, including Malin von Euler-Hogan, Carolyn O'Keefe, Morgan Moroney, Peggy Freudenthal, Deborah Jacobs, and Chris Jerome. I also owe a heaping helping of thanks to Will Staehle, who once again designed a kickass cover, and to Andrew Brozyna, who drew those delightful rebuses and illustrations of the brain.

Finally, I offer a special thanks to the many, many brainy scientists and historians who contributed to individual chapters and passages, either by fleshing out stories, helping me hunt down information, or offering their time to explain something. They're too numerous to list here, but rest assured that I haven't forgotten your help.

Notes and Miscellanea

Chapter One: The Dueling Neurosurgeons

p. 27, whether this satisfied her: Welcome to the endnotes! Whenever you see an asterisk (*) in the text, you can flip back here to find bonus material about the topic at hand, or explanations to clarify a point. If you want to flip back immediately for each note, go for it; or if you prefer, you can read all the notes at once after each chapter, like an afterword or epilogue. But do flip back: I promise plenty of exotic facts and salacious gossip. To wit:

However creepy it sounds for a duke to caress the leg of a naked fourteen-year-old princess, it sure beats the wedding night of the girl's mother, Queen Catherine. After arriving in Paris at the same age, Catherine had to consummate her marriage to Henri under the watchful eye of her new father-in-law, King François, who took a seat in the corner. François reported to advisors the next day that Henri and Catherine "both showed valor in the joust."

p. 30, marriages of art and science: After Vesalius, it became fashionable to make the drawings in anatomy textbooks as realistic as possible, to the point of absurdity. To show how faithfully they were copying every detail before their

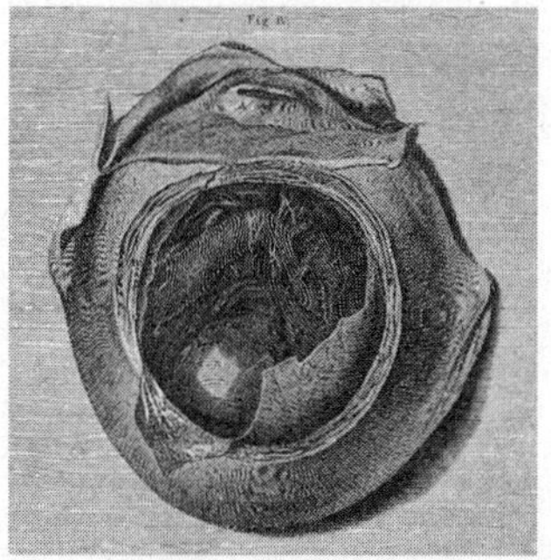

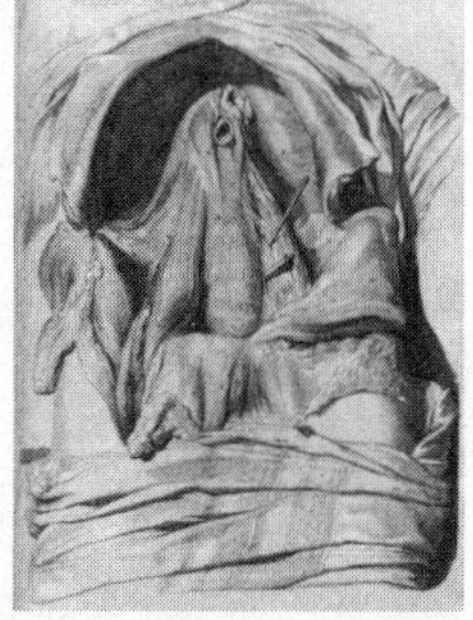

Realism taken a little too far. Left: a windowpane reflected in the unbroken amniotic sac around a fetus. Right: a fly prepares to snack on a dissected corpse.

eyes, some artists included the flies snacking on the cadaver's entrails, and one man sketched the windowpane that he saw reflected on the amniotic sac surrounding a fetus (see illustrations on previous page). A few gentlemen even bound copies of *Fabrica* and other anatomy books in human skin.

p. 35, antipodal to the blow: We don't quite know, even today, why Henri's brain suffered only a contrecoup injury. Modern studies show that the sudden acceleration of the brain—usually the result of a moving object striking the head—more often causes damage on the same side as the blow. (These are known as coup injuries.) In contrast, deceleration of the brain—the result of a moving head striking something immobile—usually causes contrecoup injuries, damage on the opposite side. But these are not hard-and-fast rules: a brain can suffer either coup or contrecoup injuries in either case—or even both, as it bounces back and forth inside the skull. Regardless, the physics of Henri's case was complicated by the fact that his head was already in motion when a moving object (the lance butt) struck it.

p. 40, macho modern athletes: Junior Seau and many other professional athletes suffered concussion after concussion during their careers, and it seems obvious to link their subsequent troubles and even suicides to this violence. We have to be careful about pinning *all* their problems on brain damage, since many retired athletes find themselves adrift for other reasons. After being told what to do every day for years, they suddenly lose structure in their lives. After being feted and pampered, they might suddenly be alone. After becoming millionaires at age twenty, they're suddenly broke. No wonder they can get depressed.

That said, there's clearly something more going on. The brains of forty-year-olds shouldn't look (as autopsies reveal) more or less identical to the brains of ninety-year-olds with neurodegenerative disease. The damage also resembles the trauma produced by IEDs in modern warfare. Military doctors have already declared chronic, widespread brain damage the "signature injury" of the recent wars in Iraq and Afghanistan. Football fans usually agonize over torn ACLs and turf toe, but brain damage may be a signature football injury, too.

The National Football League's $765 million settlement with players over concussions and brain damage is a sign that the league is finally taking this issue seriously. But the cynic in me fears that little will change, and that the money will simply soothe the collective conscience of us fans. And in some ways, focusing on NFL players misses the point. The brains of high school and college football players are still forming and are much more susceptible to damage—and they receive no financial reward for their pain. I sometimes wonder whether mothers will let their children play tackle football in ten years' time. We already have MADD. Will Mothers Against Tackle Football be next?

Chapter Two: The Assassin's Soup

p. 53, the black reaction: Another version of this legend holds that Golgi's cleaning lady deserves at least a little credit for *la reazione nera*. According to this account, one night the woman dumped both the owl brain and the silver solution into the trash, where they mingled. Golgi fished them out the next morning and decided to take a look.

p. 55, vertical organization: Neurons in the cortex generally organize themselves into six layers, and in addition to all his other discoveries, Cajal was the first person to describe this arrangement. This book doesn't delve too deeply into the workings of individual cortex layers, but if you want a quick overview, here goes.

The layers are numbered I through VI, with I nearest the scalp and VI deepest inside the brain. Data usually enter the cortex through IV, the most complicated layer. (In fact, it's often divided into sublayers—IVa, IVb, etc.) Layer IV neurons can send the information either up or down. If they send it up, layers II and III, especially, start processing it. This processing might require reaching out horizontally to other, nearby columns, but things often stay in-house. Layer IV neurons can also send information down to layers V and VI. V and VI send information to other parts of the brain, which makes sense, since they lie nearest the white matter cables that shunt information around. In general, V neurons reach out to distant parts of the brain or to the spinal cord, while VI neurons ping the all-important thalamus. (About which, more later...)

Interestingly, some scientists argue that we can tell what the brain is up to simply by noting which layers are active at any one time. If just the top layers are buzzing, that means the brain is thinking, anticipating, planning. If all six layers are active, action is likely imminent, since only the bottom layers reach into the thalamus hub and the spinal cord.

Incidentally, it's long been popular to think about the neurons in these layers as little logic gates doing computations. That's not a bad metaphor—our brains compute a lot of things—but it misses something essential about neurons. Logic gates in electric circuits are static in one sense: they do the same thing every time. Neurons aren't static. They're dynamic, and they change behavior over time, even over hours and minutes. As some scientists and philosophers have pointed out (stretching back to Plato!), a more accurate metaphor would be to think about the brain as a city and about neurons as little people. People within a city are alike in many ways—we all eat, breathe, sleep, work, complain, and so on. But we all do different activities each day, and as we mature, we change our behavior. The same goes for neurons.

p. 61, a blur of boots and rifle butts: Accounts differ about what exactly happened after Czolgosz fired the second shot—especially where people stood in line, and who knocked Czolgosz down first. I've reconstructed things as best I could. Part of the problem is that McKinley's guards changed their stories before Czolgosz's trial, perhaps embarrassed that a private citizen—and a black man—had jumped into the fray first.

Regardless, big Jim Parker became a local star. People bought scraps of clothing off his body, and he got offers on the shoes that had kicked that bastard

Czolgosz as well. Things turned somewhat sour for him later, when the guards began altering their testimony. One newspaper article—headlined "Did Jim Parker Do It?"—even floated the idea that Parker himself had pulled the trigger, not Czolgosz.

p. 64, time of death, 7:15 a.m.: On the morning of October 29, a film company run by Thomas Edison showed up outside Auburn prison to record Czolgosz's execution. Upon being turned down, they decided to restage the execution with actors a few days later. (You can see the film at http://www.youtube.com/watch?v=bZl-Z8LKSo0.) Edison did all this for the rather despicable reason that he owned patents for direct current (DC) technology and wanted to smear his business rivals, whose alternating current (AC) technology powered the electric chair.

p. 65, troubled brain was no more: After Czolgosz's death, Spitzka *fils* kept on studying how the brain changes in response to being electrocuted. In lower-profile capital cases, he often got to keep the brain, even. Unfortunately, his habit of whisking away people's brains came to the notice of organized-crime types, who resented his poaching. He started getting anonymous phone calls saying things like, *Lay off Fat Tony's brain. Otherwise, you'll regret it.* After a score of such threats, Spitzka got pretty paranoid. There are stories of him walking into a hall to give a lecture with two pistols clanging on his belt. He proceeded to check behind every door, with the guns cocked and his fingers on the triggers. He finally laid them down—and delivered a brilliant lecture on, ahem, nervous ailments. Sadly, years of continuous paranoia—coupled with a demonic work ethic—led Spitzka to start hitting the sauce pretty hard, and he became a brooding drunk. He "retired" at thirty-eight, and died at forty-six of a cerebral hemorrhage, the same ailment that killed his father.

p. 68, he had to flee: Loewi was actually arrested before fleeing. But rather than focus on, I don't know, freeing himself, his top priority was making sure his latest research results got published. He badgered his guard for a pen and paper, and then spent what could have been his last days on earth writing up a scientific article from memory. Luckily, the Nazis eventually released Loewi, albeit after he'd lost one hundred pounds in jail.

Loewi emigrated to England, then accepted a job at New York University. The only hitch was that he needed a visa first, and the U.S. embassy demanded proof that he'd actually worked as a teacher in Austria. All Loewi had was his dismissal letter from the Nazis, which wasn't exactly a glowing reference. Loewi finally found a copy of *Who's Who* and looked up his bio. The entry praised Loewi in the highest terms, and the functionary handling his case was impressed. After Loewi obtained the necessary signatures, he asked the man if he knew who'd written the entry. The man didn't. Loewi admitted he'd written it himself, then skedaddled.

Loewi's close calls didn't end there. At Ellis Island he handed over his sealed medical records to a doctor, and when the doctor opened them, Loewi read, upside down, the words "senility, not able to earn a living." In an instant Loewi saw himself being shipped back to Austria and thrown in jail to die. But the doctor had some sense, and admitted Loewi anyway.

p. 69, over one hundred different neurotransmitters: A long note, but a goodie!

The discovery of most of the hundred or so known neurotransmitters followed a similar pattern: scientists came across some new chemical in the brain, isolated it, then proved that it somehow altered the activity of neurons. The major exception to this pattern was the brain's natural painkillers, called endorphins. In this case scientists started off studying drugs like morphine and opium, which dull sensation by locking onto receptors in the brain. As the neurotransmitter doctrine emerged, scientists realized that the brain must already employ chemicals with a similar structure, or else neurons wouldn't have a receptor for morphine and opium to dock with.

The discovery of endorphins in the early 1970s was one of the messier projects in science history. A brusque young cockney lad working in Scotland, John Hughes, decided to seek endorphins—which he called "Substance X"—inside pig brains. This required Hughes to bike down to the slaughterhouse each morning before dawn with a hacksaw, hatchet, and knife in his bicycle's basket. He picked up dry ice along the way. To get his hands on the brains, Hughes had to rely on the largesse of the slaughterhouse workers who chainsawed the pigs' heads off. At first Hughes secured their cooperation by expounding on the nobility of medical research. He soon realized that scotch earned their cooperation much more quickly, and started adding a bottle to his basket. The workers brought Hughes twenty or so pig skulls each day, and while he fought off the rats, Hughes hacked out the grapefruit-sized brains in about ten minutes each, then packed them into dry ice. After returning to the lab he would pound the brains into a gray mash, then dissolve them in acetone. (Colleagues remember the lab smelling like airplane glue and rancid fat.) Finally, he'd centrifuge and evaporate off various layers, to test whether they were Substance X.

Now came the strange part. Hughes's mentor, Hans Kosterlitz, was the world's unchallenged expert on two extremely specific pieces of anatomy—the *Cavia* ileum, and the murine vas deferens, better known as the guinea pig intestines and the mouse sperm tube. Somewhere along the line Kosterlitz had determined that each of these bits—they looked like tiny, coiled worms—was superlatively sensitive to morphinelike chemicals. That is, if you suspend a *Cavia* ileum or murine vas deferens in liquid and spark a certain nerve leading into it, it will contract over and over, much like Loewi's frog hearts. But even trace amounts of morphine halt the contractions immediately. So Kosterlitz and Hughes spent months sparking the sperm tubes and intestines—producing disembodied bowel movements and orgasms in a beaker—and injecting chemical after chemical from the pig brains, to see if anything interrupted this. They finally found a substance—a yellow wax smelling of old butter—that interfered with the contractions just as morphine did. It became known as an endorphin, a portmanteau of "endogenous morphine."

By the by, drugs (illicit and otherwise) are a great way to study all the various steps involved in neurotransmission. Ecstasy, for instance, artificially floods the synapses between neurons with serotonin, allowing you to study neurotransmitter release. Cocaine prevents the vacuuming up of dopamine and other chemicals after their release. PCP, among other effects, interferes with dendrite receptors, preventing certain neurotransmitters from locking on and passing messages along. LSD lowers the ability of a neuron to inhibit its neighbors, and thereby allows sensory input to bleed from one brain region to another. Basically, for any step in the neurotransmission process, there's a drug out there that will get you high by fiddling with it.

The story about Hughes comes from *Anatomy of Scientific Discovery,* by Jeff Goldberg. You can find out more about the general history of soups and sparks in the wonderfully informative *The War of the Soups and the Sparks,* by Elliot Valenstein.

p. 71, Lee Harvey Oswald: Assassins have cut down two other U.S. presidents, Abraham Lincoln and John. F. Kennedy. Medically, both cases were straightforward—each man was as good as dead from the start. But one interesting neurological detail did arise with Kennedy in Dallas. In the Zapruder film, Kennedy famously jerks his arms up at one point, as if choking. Conspiracy-mongerers have interpreted this as evidence of his being shot from the front. But a few doctors have declared that Kennedy was actually exhibiting a primitive neurological reflex—an involuntary yanking upward of the arms in response to trauma.

By the by, Lincoln did have one interesting connection to neuroscience. As a prosecutor in the 1850s, he argued one of the first cases in U.S. history on temporary insanity. (He lost.) In the interest of space I won't go into it here, but you can read Lincoln's story online at http://samkean.com/dueling-notes. I've posted many other neurological oddities there as well, everything from why consciousness is like an old-school Nintendo to why Phineas Gage is like an android. Check it out.

Chapter Three: Wiring and Rewiring

p. 76, 250,000 miles: I can think of one other set of garments in history that possibly traveled farther than Holman's weeds: whatever unmentionables the Apollo astronauts wore beneath their spacesuits. And while people change clothes more nowadays, some modern folks of course have traveled vastly more than Holman ever could have. Hillary Clinton, as U.S. secretary of state, traveled an estimated 956,733 miles in four years, equal to two round-trips to the moon.

p. 77, segments of his optic nerves: As explained in *A Sense of the World*—a fantastic biography of Holman, by Jason Roberts—a condition called uveitis almost certainly destroyed Holman's optic nerves. But that diagnosis actually tells us a humblingly small amount, since no one knows what causes most cases of uveitis.

Holman did retain hallucinatory flashes of vision throughout his life. For instance, while chitchatting with a lady friend, a vision of what he imagined her to look like might rise before him. This proves, per the discussion at the end of the chapter, that his brain could still "see," even if his eyes couldn't. These visions delighted him for a moment but ultimately left him depressed, since they reminded him of what he'd lost.

Incidentally, there are dozens of well-documented cases, stretching back to AD 1020, of people regaining sight after decades of blindness. (Most modern cases involve corneal transplants.) You might think the most common reaction to this *Wizard of Oz* transition from dark to light would be "Wow!," but most of the newly sighted find vision kind of boring, actually, and often feel especial disappointment upon seeing the faces of loved ones. Most prefer to keep exploring objects around them through touch.

p. 78, a *superior* traveler: Blindness made Holman a superior traveler in another way: he was immune to vertigo. Whenever he joined a new ship, for instance, he would usually hand his cane to someone, remove his coat, scramble up the rigging to the top of the mainmast, then "ride" the ship like a bucking bronco. Not only did he enjoy this stunt, called skylarking, it showed his new crew that he didn't need coddling. There are other stories of Holman wandering deep inside caves and stuffing himself into enormous cannons. Perhaps most unbelievably, he tried to climb the outside of St. Peter's in the Vatican, and almost made it to the golden dome at the top.

p. 79, dictation machine called a Noctograph: Designed for writing at night, the Noctograph required no ink; Holman pressed down with a stylus onto carbon paper, which left gray traces on another sheet of paper beneath. During an era in which some men signed even bar tabs with a calligraphic flourish, the Noctograph's blocky script didn't impress: it left *t*'s uncrossed, *i*'s undotted, and *y*'s, *g*'s, and *j*'s truncated (since the guide wires made dipping below the line hard). But using it was faster and cheaper than paying someone to take dictation.

p. 83, many thousands of neurons will fire in sequence: That's just a sketch of how nerves and neurons pass information around. Since the soup/spark debates, scientists have refined their understanding of this process, so if you want to geek out, here goes:

First, nerves and neurons refuse to transmit messages unless the incoming signal reaches a certain threshold. With hearing, for instance, it's a certain volume. Otherwise, the ear hairs won't bend far enough, the nerve won't fire, and no information reaches the brain. The same general idea holds for sights, smells, and other sensory input—there's a threshold intensity. Once a cane *clack* or whatever does reach the threshold, the nerve or neuron fires. And once a neuron starts to fire, it cannot stop or hold itself back: like a gun, you can't half fire a neuron. This is called an all-or-nothing response.

What "firing" means on a micro scale is this: Once neurotransmitters lock onto a neuron's dendrites, special gates called ion channels open. This allows sodium (Na^+), potassium (K^+), calcium (Ca^{+2}), and other ions to rush into and out of the cell. The net flux of ions flips the inside of the neuron from its normal, negative state to a positive state. (This polarity flip is what the sparks detected as an electrical discharge.) This positive charge then effectively rushes down the axon to the axon tip, which finally releases neurotransmitters if appropriate. All neurons fire in this same basic way. Notice, then, that what distinguishes motor neurons from vision neurons from other neurons cannot be the *way* they fire. What distinguishes neurons—what gives them their identities—is the circuits they're wired into.

One last subtlety is that an intense noise—like that time Holman went elephant hunting in Ceylon, and rifles were sounding all around him—won't make neurons fire "harder" than a quiet noise would. Neurons always fire with the same intensity. Intense sounds merely cause the neuron to fire faster. And even this rate increase has limitations, because after a neuron has fired once, it needs to rest for a few milliseconds and recharge. If the noise increases in intensity beyond the ability of a neuron to keep up, our brains can alert us to this by firing more neurons overall.

p. 89, adult brain cannot grow new neurons: One of the most radioactively controversial topics in neuroscience in the past few decades has been whether the adult brain can in fact grow new neurons, a process called neurogenesis. Neuroscientists would once have said no, never. Today, most accept that new neurons can appear in two places: in the olfactory bulb, which processes smell, and in part of the hippocampus, which is crucial for forming memories. As for whether new neurons grow in other places, there's no consensus, to say the least.

p. 95, is a disaster: Michael Finkel wrote a fantastic profile of Kish in the March 2011 issue of *Men's Journal.* My favorite moment was when Kish made fun of Finkel's parking job, chiding Finkel for leaving his car too far from the curb. A moment later Kish popped out his prosthetic eyeballs. The article also explains that, while echolocation can revolutionize the lives of some blind people, only 10 percent ever master it.

Chapter Four: Facing Brain Damage

p. 102, provided a cover flap: Medieval and Renaissance surgeons sometimes transplanted skin from one man onto another, as an early sort of plastic surgery. But many people avoided this procedure because they believed that if the donor died before the recipient, the donor's transplanted skin flap would also die. On the other hand, those who received transplanted skin could supposedly communicate telepathically with the donor, so there was that.

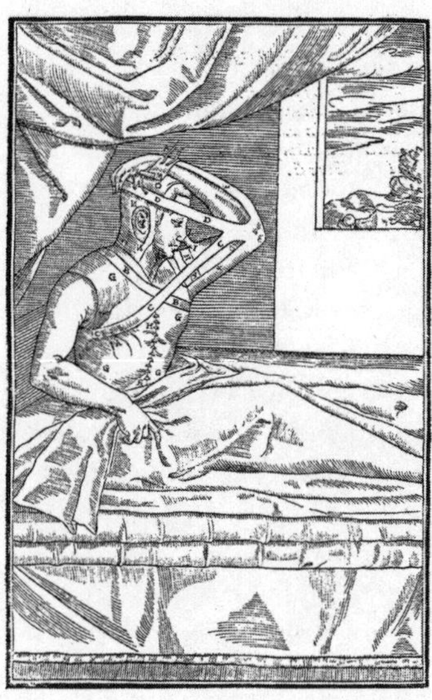

The "arm flap" method used to grow new skin, from the patient's own body, onto the nose. An early form of rhinoplasty from the 1500s.

p. 102, Francis Bacon portrait: Francis Bacon would tear out pictures of men with diseased gums and other deformities from old textbooks and use them as models for his macabre portraits, including the "screaming popes." Similarly, the designers of the blockbuster video game *BioShock* (released in 2007) dug up pictures of men with destroyed faces from a WWI plastic surgery archive and used them to create a race of mutants.

p. 106, into the left hemisphere: Note the careful wording here. It's not true that everything that your left *eye* sees ends up in the right brain. Instead, your right brain deals with everything in the left visual field—that is, everything that the left half of your left eye *and* the left half of your right eye see. Similarly, everything in the right visual field—everything that the right half of your right eye *and* the right half of your left eye see—ends up in the left brain. In other words, some visual data crosses over. (Some, but not all: because the nose gets in the way, there are slivers of sight to the far right and far left that enter only one half of the brain, a point that will become important in chapter 11, when we encounter "split-brain" patients.) Anatomically, both eyes can send information to both sides of the brain because the optic nerves that lead out of the eyes split at a point called the optic chiasm, directly beneath the brain, and certain nerve fibers cross over.

p. 112, the whole visual field: A quick note on the geometry of the primary visual cortex. I've been throwing around the word "column" because neuroscientists do, but don't take that too literally. Cells in the cortex don't form perfect stacks, like little Greek pillars. And things get even messier on a macro scale. Hypercolumns function somewhat like the compound eyes of insects, but hypercolumns don't *look* like insect eyes at all. That is, they lack the beautiful crystalline regularity of insect eyes, and they don't have regular, well-defined boundaries, either. Instead, hypercolumns look like parallel slabs of bread in some places, pinwheels in other places—and the primary visual cortex overall looks like the whorls and loops of fingerprints, with no real order or larger pattern. "Columns" and "insect eyes" are useful metaphors, but they're only metaphors.

p. 113, *a Corvette*: Following this logic further, you might expect neurons in the what stream to get more and more specific in what they respond to at each step—until you eventually reach one single neuron that lights up, lightbulb-style, only for a certain Corvette-with-dice-in-the-mirror-and-inappropriate-bumper-sticker-that-your-creepy-neighbor-owns. Some neuroscientists did once believe in this step-by-step convergence to a single neuron, which they called a "grandmother cell," since you'd have to have one dedicated just to her. But the idea now stands in low repute. Again, instead of a single neuron flickering on, the brain almost certainly looks for *patterns* of neurons that flicker on en masse. You don't have grandmother cells, but probably do have "grandmother ensembles."

p. 114, important note: If you forget everything else in this book—the title, my name, all the sex and violence—please, please remember this: *nothing in the brain is strictly localized.* Everything your brain does depends on many different parts working together—there's no "language spot," no "memory spot," no "fear spot," no (heaven help us) "God spot."

Now, it's true that some parts of the brain do play a bigger role in language or whatever than other parts. And people do sometimes, as a shorthand, refer to a spot "for" some faculty. (I do it, too!) But talking about one specific spot like that is a deliberate oversimplification.

And be hyperskeptical of news stories that purport to show, with candy-colored brain scans, an anatomical island that "explains" some complicated attribute of the human mind. It ain't never so simple. Some neuroscientists have criticized the worst brain-scan studies as "brain porn," "brain voodoo," and "the new phrenology."

p. 119, the hang of humans again: This story about the shepherd brings up an interesting question: if this man could recognize individual sheep by looks, could the sheep recognize him? Probably not. Sheep seem to lack the cognitive circuitry. But at least one animal, the crow, can tell human faces apart. Science reporter and radio host extraordinaire Robert Krulwich once did a story about a biologist who harassed crows in the course of his research—to the point that the crows would dive-bomb him (and only him) whenever he passed by. The biologist wondered how they picked him out, and he concluded through a series of experiments that they could recognize his face.

First, he donned a caveman mask and harassed a murder of crows until they learned to hate the masked figure. He then transferred the mask to people of all different shapes, sizes, and gaits—old people, children, limpers, bald people. The crows promptly transferred their loathing to the person with the mask. The kicker was that when he put the mask on upside down, the crows would swoop by upside down to get a look at him. You can hear the story at http://www.npr.org/ blogs/krulwich/2009/07/27/106826971/the-crow-paradox.

p. 123, hand gestures in prelanguage communication: Funnily enough, the neuroscientist who discovered hand-loving neurons, Charles Gross, discovered them in a way quite similar to Hubel and Wiesel's discovery of line-loving neurons. It was 1969, and Gross had wasted hours one night trying to get certain neurons inside a spider-monkey's visual cortex to respond to something—anything. Desperate, he waved his hand in front of a projection screen near the monkey's eyes, as if to say, *Pay attention, damn it.* The neuron did: rat-a-tat-tat-tat-tat-tat. Gross spent the next twelve hours playing shadow puppets and cutting out shapes and holding them up to determine which outlines the neuron liked best. The answer? Slender monkey hands, natch—hands with longer and slimmer fingers than human hands.

While we're speaking of Gross, I highly recommend his books *A Hole in the Head* and *Brain, Vision, Memory*.

p. 125, "Mr. Potato-Head without the features": The quotes here and below, as well as many more details about Dallas Wiens, can be found in Raffi Khatchadourian's excellent story in the February 13, 2012, *New Yorker*.

Chapter Five: The Brain's Motor

p. 136, the pineal gland: In truth, the pineal gland—a remnant of a third eye that vertebrates used to have (really)—helps detect light and influences our sleep/wake cycle. And yes, because I know you were thinking it, the name of the pineal gland does seem to derive from another body part, since a few early anatomists insisted that it looked just like a penis. That's far from the only neurostructure that had a ribald name way back when. Gutter-minded anatomists also named various brain bits after the buttocks, testicles, vulva, and anus.

p. 141, ringed every hospital: In these days before antiseptics, doctors themselves also suffered high mortality rates. Florence Nightingale, a nurse during the Crimean War (1853–1856), watched one particularly inept surgeon cut both himself and, somehow, a bystander while blundering about during an amputation. Both men contracted an infection and died, as did the patient. Nightingale commented that it was the only surgery she'd ever seen with 300 percent mortality.

p. 144, tied up with phantoms: No one knows whether the rise in male-to-female sex-change operations in the past few decades has led to a corresponding rise in the number of cases of phantom penis, but perhaps not. Scientists who study phantoms have noted that most male-to-female transsexuals never felt their genitals belonged to them anyway—possibly because their hardwired, internal body scaffolds were anatomically female. If that's true, phantoms wouldn't appear. Along these same lines, just as people born without arms or legs can feel phantom limbs, these scientists predict that female-to-male transsexuals should feel phantom penises from an early age.

p. 147, greater pleasure: Neuroscientist V. S. Ramachandran has suggested that one common sexual fetish, a foot fetish, might result from cross-wiring between the brain map's foot areas and genital areas. He admits he's speculating here, but argues that this idea is at least as plausible as Freud's explanation—that to our nearsighted subconscious, the foot resembles a penis. By the way, Ramachandran is one of the most brilliant and creative neuroscientists out there, and I highly recommend his books *The Tell-Tale Brain* and *Phantoms in the Brain.*

p. 152, take phantom limbs seriously: Just to be clear, Mitchell did not believe in spiritualism and intended the ending of "The Case of George Dedlow" to be a farce. He loved exposing mediums as frauds, in fact, and was bemused when spiritualists seized upon the story as "proof" that séances worked. Also, it's not clear why people clamored to visit Dedlow in Stump Hospital, since the story clearly (albeit briefly) states that he got transferred to another hospital later.

And in case you're wondering, the U.S. Army Medical Museum (now the National Museum of Health and Medicine) does have two specimens numbered 3486 and 3487, but they're not legs. The first is a cranium fragment from an Illinois private wounded near Atlanta; the second is a left humeral fragment from a Michigan private also wounded near Atlanta. The latter survived.

Chapter Six: The Laughing Disease

p. 165, mini-brain all by itself: Odd fact: the cerebellum contains something like three-quarters of all the neurons in the brain. There are two possible ways to interpret this. One, despite what we might think—that movement is a "low-level" brain function, and cognition a "high-level" brain function—movement actually requires plenty of sophisticated brainware. Two, perhaps the cerebellum plays a bigger role in cognition than scientists traditionally give it credit for.

p. 174, barbaric "bushman" stereotypes: Gajdusek always felt protective of the Fore. Besides defending them against "bushman" stereotypes, he seethed when an Australian tabloid dubbed kuru "the laughing disease," which he found derogatory and flip. On the other hand, when Gajdusek wanted to shock people, he was not above indulging in stereotypes himself. In a letter to his mother he once bragged, "[My hosts] were still spearing each other as of a few days ago."

p. 178, dead neurons and spongy holes: No one knows whether prions kill neurons directly or indirectly. Perhaps the buildup of prion plaques simply creates a toxic by-product, or disrupts some key process in a roundabout way. But it's clear that there's a strong correlation between prion plaques and neuron damage. One complicating factor is that scientists still don't know what the normal, healthy prion protein does. Different experiments have linked it to the production of myelin sheaths, the birth of new brain cells, the plastic rewiring of circuits (especially in young brains), and the transport of copper ions. And while it's especially active in brain cells, all cells seem to manufacture it.

By the by, Kurt Vonnegut fans might have noticed an analogy between prions and the "ice-nine" of *Cat's Cradle,* a special form of ice that turns solid at room temperature and that creates more of itself by latching onto and corrupting other water molecules. Prusiner had read the book and enjoyed drawing the comparison.

p. 180, in subsequent interviews: Those who knew and loved Gajdusek continue to debate his guilt or innocence. Many details of his life look like classic signs of pedophilia, not least his decision to go into pediatrics. On the other hand, he supposedly pled guilty to the molestation charges only to avoid a lengthy trial that would have bankrupted him. Friends also alleged that the FBI promised to support the first accuser financially if he accused Gajdusek—which doesn't invalidate his testimony but does make it more suspect. That the allegations emerged in the 1990s—when the U.S. went hysterical over child abuse—raises suspicions as well. And lord knows that cops have railroaded people into false confessions before.

That said, at least four allegations of sexual abuse and inappropriate touching emerged against Gajdusek before the FBI got involved; all were dismissed for lack of evidence, but the pattern looks disturbing. Plus, in the heartbreaking BBC documentary *The Genius and the Boys,* Gajdusek claimed on camera to have had sexual contact with hundreds of boys across the world. That's probably an exaggeration: Gajdusek loved provoking people and mugging for the camera. (At one point in the movie he also speaks out in favor of "intergenerational" incest, and says that children should enter their parents' bedrooms at night and assist with sex.) But the documentary claims that at least seven men have now accused Gajdusek of having sexual contact with them as boys, and one, an American, does so on camera in the film.

Chapter Seven: Sex and Punishment

p. 188, a morphine addict: Society had a higher tolerance for addicts back then, especially among gentleman professionals. The morphine junkie, William Halsted, also had a taste for cocaine. And William Sharpe told a disturbing story about seeing one eminent surgeon gulp a slug of whiskey right before an operation. Five minutes later the surgeon tried to punch a hole in the back of a patient's skull to access the brain, but he pushed too hard, puncturing the brainstem and killing the patient instantly. He muttered to a few visiting doctors, "It seems that the operation is over," then slunk off. Sharpe found him in the doctors' changing room drinking more whiskey, his hands shaking.

p. 191, hands, feet, and facial bones thicken: Synthetic growth hormone makes an excellent performance-enhancing drug, and professional athletes who dope with it often develop a form of acromegaly. In short, their skulls and jawbones swell, and their hat and helmet sizes increase markedly.

p. 194, his two thousandth brain tumor: Eleven of those two thousand tumors came from the noggin of one man, Tim Donovan. Cushing eventually removed 47 ounces of cancerous tissue from Donovan's skull—the equivalent of a full extra brain.

p. 197, pings the hypothalamus: On March 29, 1916, a Mr. O. in north-central Iowa shot himself through the right temple. Someone discovered him a dozen hours later, and a doctor revived him at a local hospital. The bullet severed O.'s optic nerves and he woke up blind, but the complication that made his case memorable didn't start until a few days later. During the first two days of convalescence O. urinated just 14 ounces total (0.1 gallons). On day five the floodgates opened, and he began peeing, and peeing—and peeing. By April 10 he was peeing 1.2 gallons daily, three times what an average male should. And that doesn't include the three times he wet his bed. On April 12 he peed 1.5 gallons, again not counting accidents.

His doctor eventually traced this Niagara to the hypothalamus, which scientists now know manufactures a neurotransmitter called vasopressin. Vasopressin, after seeping into the bloodstream, turns on pumps in the kidneys that reabsorb H_2O and keep it out of the bladder. O.'s shot hadn't hit the hypothalamus directly, but brain tissue along the bullet's path swelled and slowly crushed it. When vasopressin production shut down, his rapidly filling bladder had no choice but bail and bail and bail some more. O. spent the rest of his sad life in an asylum.

As another example of the limbic circuit interacting with the wider body, consider the case of Rita Hoefling, a forty-year-old white South African housewife. Surgeons removed Hoefling's adrenal glands in the early 1970s after diagnosing her with Cushing's syndrome, which occurs when the adrenals release too much cortisol. The surgery stopped that problem but stirred up other trouble. The adrenals check the activity of the pituitary gland, and with nothing holding the pituitary back now, it began to churn out hormones that increase the production of melanin inside skin cells. Melanin changes the color of skin, and Hoefling began to turn bronze, then light brown, as a result. This well-known side effect of removing the adrenals (Nelson's syndrome) wouldn't have caused much of a stir—except in apartheid South Africa. Hoefling started getting thrown off whites-only buses. Her husband and son abandoned her. She was even barred from her father's funeral. After her ostracism, the colored community magnanimously embraced Hoefling, and she later spoke out against the evils of apartheid.

p. 201, yawns: Yawns are a rich neuroscientific subject. As with smiles, some people who are paralyzed due to a stroke in the motor area can still reach out and stretch their arms if induced to yawn. That's because the yawn reflex originates in the brainstem and can therefore bypass the damaged neurons and access the arm muscles through different channels. The origin in the brainstem also implies that yawning is an ancient reflex, far older than humanity. And sure enough, many other mammals yawn, as do snakes, birds, and turtles. Humans born with only a brainstem (i.e., without the higher parts of the brain) can also yawn, as can fetuses in the womb.

Only two animals, chimps and humans, can "catch" yawns, though. And humans don't pick up contagious yawning until age four or five, implying that we need to develop certain parts of the brain first, probably those related to social skills and empathy. (Along these lines, people with autism don't pick up contagious yawning until much later in life, if at all.) Furthermore, we don't catch yawns equally from all people: loved ones infect us with yawns more readily than good friends, who infect us more readily than acquaintances, who infect us more readily than strangers. This leads me to wonder whether you could tell someone was falling out of love with you by timing their yawn delay.

Finally, the big question is, and always has been, why we yawn at all. And the answer is, and may always be, that no one knows.

p. 208, Damasio's work here: Damasio discusses Elliot in detail in his book *Descartes' Error.* Check it out for a full understanding of all the subtleties of the case.

Chapter Eight: The Sacred Disease

p. 220, remain conscious during surgery: Lobotomist Walter Freeman loved to recount a story from early in his career, when he asked a surgical patient what was going through his mind at that moment. "A knife," the man replied. There's also an incredible video out there of bluegrass musician Eddie Adcock. To ensure that nothing was wrong with his brain during surgery, rather than have Adcock talk, his surgeons let him play his banjo. You can see footage at http://news.bbc.co.uk/2/hi/science/nature/7665747.stm.

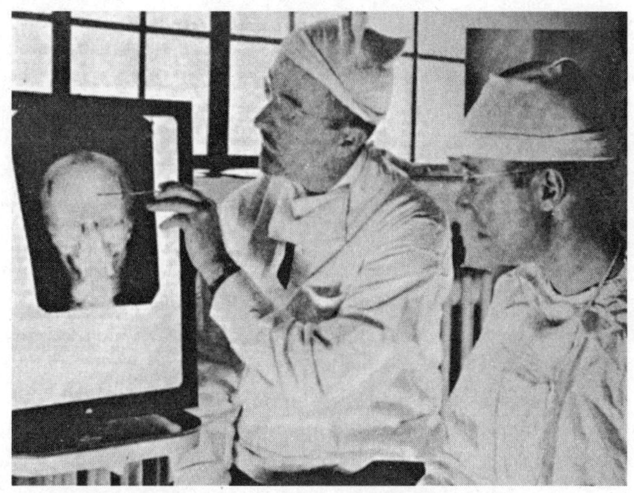

Walter Freeman, the notorious American lobotomist.

p. 224, ancient Egyptian priests: Ancient Egyptian priests may have discounted the brain, but at least some Egyptian doctors did not. The so-called Edwin Smith papyrus—based on material first compiled perhaps five thousand years ago, before many of the pyramids were built—outlines treatment options and the probable prognoses for dozens of types of head and brain injuries. For the first time in history the document also singles out the brain as a distinct organ instead of just a general part of the head. The hieroglyphic translation of "brain" in the papyrus means "marrow of the head."

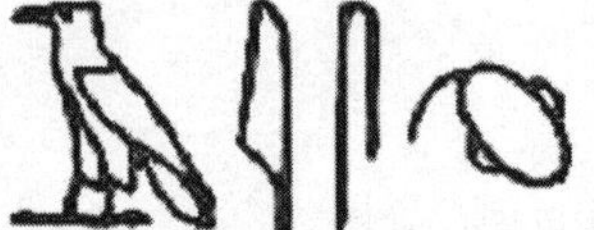

The Egyptian hieroglyph for "brain." Literally, it means "marrow of the head."

p. 235, a common occurrence in the 1920s: Although Cushing drastically lowered the risks associated with neurosurgery in his own practice, other surgeons were slow to adopt his ways, and it remained one of the deadliest procedures out there. As late as 1900, some 75 percent of neurosurgery patients in London hospitals died of complications. (Primitive trepanners working in New Guinea at this time, who opened skulls with shark teeth and dressed wounds with banana leaves, killed just 30 percent of patients, which really makes you think.) In addition to antiseptics, work on brain localization helped reduce mortality a lot: rather than simply crack the skull open and blunder about for a tumor or scar, surgeons could now study their patients' deficits and make an educated guess about where to open them up.

Chapter Nine: "Sleights of Mind"

p. 245, vacuum of power: There's still no clear guideline about who, if anyone, can judge whether the president is mentally capable of executing his office. We often say the vice president is one heartbeat away from the presidency, but the country itself—especially given the delusions that often beset stroke victims—is always one little arterial tear away from a constitutional crisis.

p. 253, electrical conductivity of skin: Lie detector tests also measure the electrical conductivity of the skin (as well as heart rate, blood pressure, breathing patterns, and other signs of physiological stress). That may sound odd, since polygraphs are often denounced as pseudoscientific hooey, but there's a difference between the way scientists use them and the way your typical asshole interrogator does. The interrogator claims that just because you're showing signs of nervousness, it means you're lying—when of course you could be nervous for dozens of reasons, not least because he's threatening you with jail. Scientists use these tests simply to measure whether or not you're experiencing an emotion, period, and whether those signs change from stimulus to stimulus. Scientists don't (or at least shouldn't) claim to be reading your actual thoughts based on such signs, and in this more restricted case, the skin conductivity test does provide meaningful information.

As a fun sidenote, the first crude polygraph machine, developed by a Harvard undergraduate named William Moulton Marston, consisted mainly of tubes that wrapped around the person being tested like a boa constrictor. Marston later (under a pen name) went on to write comic books, and he invented Wonder Woman—who of course wielded a golden "lasso of truth" that, when it ensnared bad guys, forced them to be honest and forthright. Unfortunately, Marston wasn't always so honest and forthright in his own career: he used the polygraph in advertising research and got caught making up data when trying to determine whether young men preferred one brand of disposable razor over another.

p. 255, a safe, familiar context: Some people experience déjà vu constantly, as if everything that's happening to them at that moment has already happened to them before. This can lead to some unintentionally funny complaints. Some victims refuse to watch television since everything seems like a rerun. One woman gave up her library card because she'd already read every last book there. Another woman quit playing tennis because she knew the outcome of every point beforehand. One man even claimed, against all logic, to have attended a certain funeral many years prior. Scientists don't know why déjà vu occurs, although there are plenty of guesses. One good guess is that memories run through the brain like a loop of video. Normally we record the material first and then replay it later, but if for some reason you recorded the memory and started playing it back immediately, you'd experience déjà vu. Other scientists attribute déjà vu to different causes, and of course there could well be multiple causes.

Incidentally, in case you're wondering, the hearing-limbic circuit can also crap out and disconnect someone's voice from its emotional glow. Indeed, there are cases of blind people suffering from auditory Capgras delusions. The limbic-touch and limbic-smell circuits can also die: a blind Brazilian woman who suffered from Capgras complained that her husband's double felt fatter and smelled different.

p. 261, *Alice* victims: If you want to go there, neuroscientists have pointed out that *Alice in Wonderland* is full of potentially interesting neurological cases. Humpty Dumpty cannot recognize faces (face-blindness), and he suffers a catastrophic brain injury after falling. The dormouse at the tea party has narcolepsy. The White Queen has dyscalculia, the inability to do arithmetic ("I can't do subtraction under any circumstances," she says). And hosts of other characters betray bizarre beliefs about space, time, and the nature of existence.

Another delusion that seems to have sprung from the pages of *Alice* is the glass delusion, wherein people believe they're made of glass. Oddly, sufferers often thought of themselves as specific items, such as urinals or oil lamps. A surprising number also believed that they had glass buttocks, including Charles VI of France, who wore reinforced clothes to protect his glass bum. In a variation on this theme, Princess Alexandra of Bavaria insisted that she'd swallowed a glass piano, and that it remained intact inside her.

p. 264, if that's true: While the results of Libet's "free will" experiments are robust, the interpretation of them remains disputed, to put it mildly. Some scientists and philosophers argue that no matter how good someone's reflexes are, there will always be a lag between when he decided to move and when his eyes recorded the time on the clock. Others object to the idea that you can pin down consciousness so precisely, to a specific millisecond of activity; perhaps consciousness is "smeared" more widely across time. Or perhaps we really are zombies when it comes to crude motor tasks like raising a finger, but we have free will with bigger, more consequential decisions. Along those same lines, perhaps our free will "programs" our brains ahead of time. So even though we spend most days doing most things on autopilot, the habits we've set up were indeed freely chosen.

Libet himself found most of those objections either baseless or spurious. Still, he didn't believe he'd wiped out free will, not entirely. Libet did accept that our conscious selves do not have the free, unfettered ability to initiate action: that talent belongs to the unconscious. However, he argued that people did have the choice—the real, free choice—to squelch those unconscious impulses and refuse to act on them. As he put it, we don't have free will, but we might have "free won't." The window for clamping down on the unconscious's decisions was short—just 150 milliseconds—but it would make people morally responsible for their actions. And as Libet once wrote, "Most of the Ten Commandments are 'do not' orders." It's not the free will most of us believe in, but it might be all that neuroscience leaves room for.

Chapter Ten: Honest Lying

p. 275, groundbreaking work in 1947: Amazingly, Hugh de Wardener is still alive and living in England, and he generously agreed to be interviewed for this book.

p. 278, the American lobotomy bandwagon: As mentioned, lobotomies sever the white matter connections between the frontal lobes and limbic system. And although lobotomies stemmed from the work of a Portuguese doctor, who won a Nobel Prize for it, the procedure really took off in the United States, which has always embraced quack medicine and quick fixes. In all, approximately fifty thousand Americans underwent lobotomies between the mid-1930s and late 1950s, with one single doctor, Walter Freeman, doing three thousand by himself. Freeman roamed the hills and byways in a mobile clinic dubbed the Lobotomobile, and performed the procedure with a rubber mallet and an ice pick from his kitchen. He operated on children as young as four, and his record was twenty-five lobotomies in one day. His most famous patient was Rosemary Kennedy, sister of JFK, who spent the remainder of her life in an institution.

p. 288, personal knowledge: In a normal brain, all knowledge is probably episodic at first and therefore relies on the hippocampus; it becomes semantic knowledge only later, after it works its way free of the hippocampal web and becomes context-free. For instance, you probably first learned that Abraham Lincoln was the sixteenth U.S. president on, say, a field trip to Washington, D.C., or (more likely) when you got that item wrong on a quiz. But eventually you forgot the specific moment of learning and retained only the more abstract knowledge that Abe = 16.

The episodic/semantic distinction sheds light on confabulators as well, since they primarily lie about personal episodes. Looking ahead, confabulators also usually feel both familiarity and recollection for even their most Münchhausenian tales. Indeed, that's their problem in a nutshell.

p. 290, divides up responsibility: This is just a plausible sketch of how memory works inside the brain, and plenty of neuroscientists disagree with the details presented here. In other words, like pretty much everything in neuroscience right now, it's open to revision. Just saying.

p. 292, "no distinct limits" to Shereshevsky's memory: Shereshevsky's finite brain did not have an infinite capacity for storing information, of course. But when considering claims like this, keep in mind that the way we commonly think of memory—as a "jar" or "hard drive" or something else that can fill up—is misleading. As some scientists note, it's better to think about memory as a muscle—a faculty that, if exercised, gets stronger and stronger. So Shereshevsky's continual acquisition of new material wouldn't necessarily push the old material out.

p. 296, we do garble details: Probably the best example of people garbling a personal memory involves September 11, 2001. Survey after survey has revealed that people have explicit memories of watching television news coverage of the two planes hitting the World Trade Center towers on that day. But that didn't happen. No television news station broadcast the impact footage until the next day.

Chapter Eleven: Left, Right, and Center

p. 305, this location: To be specific, Broca localized the lesions in Tan's and Lelo's brains to the third convolution of the frontal lobe, near where the frontal lobe and parietal lobe intersect. This region eventually became known as Broca's area.

Although Broca preserved the brains of both Tan and Lelo for later generations, later generations almost lost them. Twice. Before dying, Broca deposited the brains in the Musée Dupuytren, a museum located in the dining hall of a former monastery. The Dupuytren's walls collapsed during a bombing raid in 1940, and during the move to a more permanent home, the brains disappeared. Not until 1962 did a scholar hunt them down. They promptly disappeared again, when a janitor who'd moved them up and died without telling anyone where he'd moved them. But they turned up again in 1979, and remain (for now) safe. For some reason Broca bottled Tan's brain vertically, so it rests on the frontal lobes.

p. 308, first and second languages: Whether English, Tagalog, or Serbo-Croatian, the first language you learned was stored in your procedural memory, which explains why speaking it comes so naturally: it's subconscious. With a second language, the situation varies. If you learn a second language "naturally" (i.e., in daily life), it will also enter procedural memory and will become nearly automatic, especially if learned when young. If you acquire a second language through formal tests and schooling, it enters declarative memory and doesn't stick as readily. This distinction helps explain why bilingual people can lose either language, since different memory systems can suffer damage independently. (By the by, the reason bilingual people usually swear and coo to babies in their first language is that first languages, being subconscious, are more deeply intertwined with our emotions.)

One odd disorder related to multilingualism is "foreign accent syndrome," which occurs when people wake up from a stroke or head trauma and suddenly speak with an accent. One Englishwoman, for instance, woke up sounding like a parody of a French lady, zaying all zhortz of fun-nay zhings. The disorder sounds dramatic but actually has a prosaic explanation. It turns out that the trauma simply reduces the "acoustical spectrum" in someone's brain. As a result, her teeth, tongue, and lips cannot make all the sounds she needs to, and for whatever reason, other people interpret her limited range as a stereotypically foreign accent.

p. 309, Auburtin: Auburtin's father-in-law—a man named Jean-Baptiste Bouillaud—had offered 500 francs for any proof of a widespread lesion in the frontal lobes without attendant loss of speech. Bouillaud eventually lost this bet, albeit under dubious circumstances. At a Société meeting in 1865, Dr. Alfred Velpeau recounted the case of a sixty-year-old wigmaker admitted to his care some years before for excessive incontinence of urine. Apparently no man had ever blathered as much as this patient did: the wigmaker talked incessantly, compulsively, prattling on even while he slept. Nothing could shut him up. He died shortly afterward, and although Velpeau hadn't planned to examine the brain during the autopsy, he decided to do so at the last minute. Lo and behold, he found that a tumor had destroyed the man's frontal lobes. At least that's what Velpeau claimed. Because the wigmaker had died in 1843 and Velpeau stepped forward only a quarter century later, some people suspected him of fraud. There was a row at the Société meeting where he claimed the prize, but Bouillaud eventually paid up. (From a modern perspective, the wigmaker probably did suffer some sort of frontal lobe damage, which would have lowered his inhibitions and caused him to blather. But the frontal lobes are quite large, and quite a lot can get damaged in them without affecting Broca's area.)

p. 317, killed Lashley's theory: From today's perspective, Lashley's ideas aren't total bunk. Language and memory and other complex faculties do draw on multiple parts of the brain. But brain signals don't get spirited around via electric waves; they're carried by ions and chemicals. And saying that multiple parts of the brain contribute to something is a far cry from Lashley's claim that all parts of the brain contribute equally.

As for Lashley's experiments, rats could still navigate mazes after suffering brain damage because rats, the tricky bastards, have several ways of finding their way around—touch, smell, hearing, sight. Rats even have different vision centers in different parts of the brain. The lesions no doubt impaired these systems, but you'd have to knock them out completely to render a rat helpless. This is one reason why rats will still be around long after human beings have perished from the face of the earth.

p. 318, it would start getting lost again: Sharp-eyed readers may have noticed that the cat/maze/eye patch experiment wouldn't have worked as described here, since each eye provides some input to both halves of the brain. (Again, it's input from the left and right visual field that ends up in the right and left brain, not input from the left and right eye per se.) Sperry knew this, of course, and when he cut the cats' corpus callosums, he also surgically rewired their optic nerves, so that the nerves provided input to one hemisphere only. As is probably obvious, Sperry was a gifted surgeon: the hand-eye coordination he developed on the playing fields of Oberlin served him well.

p. 319, draw a picture: Forcing split-brain people to draw with their weaker hands (usually the left) produced some pretty crappy art, even by the generous standards of neuroscientific testing (see below), but it was important to isolate the abilities of the left and right brain. Interestingly, though, the left hand of split-brain people was artistically superior in some ways. That is, the lines that the left hand drew were wobbly, because of lack of practice; but overall, the lefty picture did resemble what it was supposed to, since the right brain has good spatial skills. In contrast, the lines drawn by the right hand were sure and firm—but the overall depiction looked terrible, because the left hemisphere lacks a sense of space.

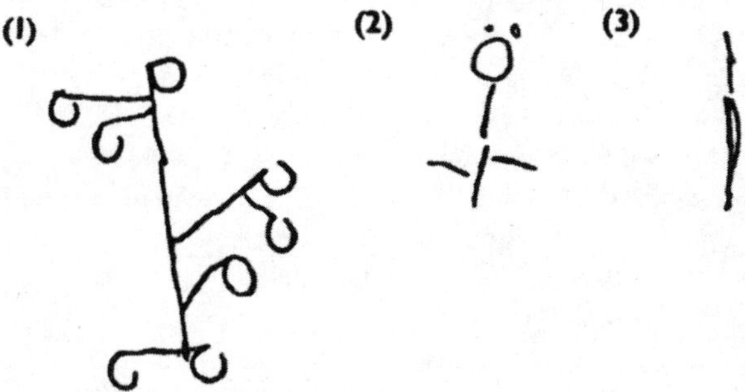

Pictures drawn by someone with severe visual agnosia in the brain. (1) A tree. (2) A man. (3) A boat. Notice that the man's eyes appear outside his head.

p. 321–322, animals show subtle hemispheric differences: In contrast to most animals, nature herself is not oblivious to left-right differences, especially not on small scales. Some subatomic particles come in left- and right-handed varieties, and one of the fundamental forces of nature (the weak nuclear force) interacts with each version differently. Even more important, all known life on earth uses DNA that forms a right-handed spiral. (Point your right thumb toward the ceiling; DNA twists upward along the counterclockwise curl of your fingers.) Left-handed DNA would actually kill our cells, and yet biology textbooks can go through multiple editions showing "backward" DNA, without anyone noticing. Not that I should talk: the cover of my second book, *The Violinist's Thumb,* shows a backward DNA strand, which I didn't notice until an eagle-eyed reader pointed it out.

p. 324, Perhaps those aren't just metaphors: Speaking of metaphors, there's strong evidence that whenever we hear or read an action verb (run, hit, bounce)—or even when we use certain metaphors (he swallowed his pride, she juggled two careers, it stretched our understanding)—our motor centers start humming in response. Not enough to move the body, but something's afoot. Apparently this stimulated physical activity helps our minds, well, grasp the concept. In this and other ways, much of language is literally embodied. For more on this subject, see the book *Louder Than Words*, by Benjamin Bergen.

Chapter Twelve: The Man, the Myth, the Legend

p. 337, a process in a population: If you're having trouble wrapping your head around the idea of consciousness not being a thing in a place but a process in a population, consider this wonderful analogy from V. S. Ramachandran. In his book *Phantoms in the Brain,* he considers an episode of *Baywatch* and then asks where exactly the episode is located. On the beach where the actors were filmed? In the camera that recorded the drama? In the cables pumping bits into your television? In the television itself? (And if so, where in the television—its electronic bowels, the LCD screen?) Perhaps the show is located in the storm of photons arriving at your eyes? Perhaps in your brain itself?

After a few seconds it becomes clear that the question doesn't make sense. Or rather, it misses the point. The real question isn't where the episode is located but how the various pieces of technology transmit a moving picture across time and space and into your brain. Similarly, Ramachandran suspects that as we learn more about how the brain produces consciousness, we'll care less and less about specific locations.

p. 342, remaining cognizant: The delightfully gruesome paper "Transcranial Brain Injuries Caused by Metal Rods or Pipes Over the Past 150 Years" covers a dozen cases of people whose skulls were impaled by metal objects, and in five of the twelve, the victims didn't black out for even a moment. Two memorable cases include a drunken bow-and-arrow game called "William Tell" and an assembly-line accident in which a twenty-six-foot metal shaft plowed most of the way through a man's skull before getting stuck. Because he never passed out, the man could feel it sliding through his head inch by inch.

More anecdotally, during a domestic dispute in Mississippi in 2009, a woman got shot in the forehead with a .38. The bullet passed clean through her brain, front to back. And not only did she remain conscious, she remembered her manners: when a police officer knocked on her front door minutes later, he found her making tea, oblivious to her injury, and she insisted he take some.

p. 343, to be torn out again: Clive Wearing will never recover, sadly. That said, his amnesia was most acute during the first decade after his illness, and there's some evidence that his symptoms have abated since about the year 2000—probably due to plastic changes in his brain that allowed him to recover some function. Neuroscientists have not yet documented this improvement with proper studies, so we have to be cautious. But Wearing's wife, Deborah, who spends more time with him than anyone else, insists that Wearing has gotten better.

For instance, Clive's memory has improved to the point that he can have meaningful, if brief, conversations with Deborah, rather than just repeat the same things over and over. And while he still "pops awake" repeatedly, he's gotten used to the epiphany now after millions of times, and no longer records it so zealously. He can even follow certain films a little (e.g., James Bond movies), and can be out in public without wandering off. Deborah discusses these and other improvements toward the end of her heartbreaking memoir *Forever Today.*

p. 344, will never abandon him: There's one seeming exception to the rule that not even amnesiacs lose their sense of self. People who suffer from so-called memory fugues do seem to forget their personal identities: they're more like amnesiacs on television shows, who wake up not knowing anything about their past lives. But even fugue victims retain something from their past: they might be able to log on to an e-mail account, for example, because of muscle memory. And fugue victims usually do assume a new identity, since the brain apparently cannot function without a sense of self.

I've written up a bonus story about the most famous fugue victim in history, an American farmer named Ansel Bourne. You can read it online at http://samkean.com/dueling-notes.

p. 349, some modern historians: The historian who deserves the most credit for revising our view of Gage—and for demonstrating that Gage might well have recovered some skills and functions later in life—is Malcolm Macmillan, author of the delightful book *An Odd Kind of Fame.* Everyone who cites Gage's case should read Macmillan first—he deserves a ton of props for taking on a popular but inaccurate legend. Macmillan also suggests that Gage's story is worth remembering because "it illustrates how easily a small stock of facts can be transformed into popular and scientific myth." Wise words.

As long as we're talking about inaccuracies, I should note that, obviously, I've had to simplify the Gage story and leave out some details. For instance, another doctor besides John Harlow did examine Gage a year after the accident—one Henry Bigelow, who provided important additional facts. I focused on Harlow's account instead of Bigelow's mainly because Harlow alone discusses Gage's mental functions. See *An Odd Kind of Fame* for the full story.

That said, I can't resist including some biographical details about Bigelow, who had, shall we say, a colorful youth. As one historian noted, Bigelow is remembered today as a "heavily bewhiskered surgical giant" who entered Harvard Medical School at age fifteen. But while attending Harvard, Bigelow spent most of his time "making loud noises, joining drinking clubs... and manufacturing nitrous oxide for the customary annual binges of the chemistry class." Bigelow was finally expelled from Harvard for conducting "pistol practice in his dormitory room," a stunt that also got him "banned from the town of Cambridge for the remainder of the year. But despite his rustication he managed to graduate on schedule."

Works Cited

General

Albright, Thomas D., et al. "Neural Science: A century of progress and the mysteries that remain." *Cell* 100, no. 25 (2000): S1–55.

Bergen, Benjamin K. *Louder than words: the new science of how the mind makes meaning.* New York: Basic Books, 2012.

Bor, Daniel. *The ravenous brain: how the new science of consciousness explains our insatiable search for meaning.* New York: Basic Books, 2012.

Doidge, Norman. *The brain that changes itself: stories of personal triumph from the frontiers of brain science.* New York: Viking, 2007.

Feinberg, Todd E. *Altered egos: how the brain creates the self.* Oxford: Oxford University Press, 2001.

Finger, Stanley. *Origins of neuroscience: a history of explorations into brain function.* New York: Oxford University Press, 1994.

Gazzaniga, Michael S., Richard B. Ivry, and G. R. Mangun. *Cognitive neuroscience: the biology of the mind.* New York: W. W. Norton, 1998.

Goldstein, E. Bruce. *Sensation and perception.* Belmont, Calif.: Wadsworth, 1989.

Gross, Charles G. *Brain, vision, memory: tales in the history of neuroscience.* Cambridge, Mass.: MIT Press, 1998.

———. *A hole in the head: more tales in the history of neuroscience.* Cambridge, Mass.: MIT Press, 2009.

Harris, Sam. *Free will.* New York: Free Press, 2012.

Klein, Stephen B., and B. Michael Thorne. *Biological psychology.* New York: Worth, 2006.

Macmillan, Malcolm. *An odd kind of fame: stories of Phineas Gage.* Cambridge, Mass.: MIT Press, 2000.

Magoun, Horace Winchell, and Louise H. Marshall. *American neuroscience in the twentieth century: confluence of the neural, behavioral, and communicative streams.* Lisse, Netherlands: A. A. Balkema, 2003.

Ramachandran, V. S., and Sandra Blakeslee. *Phantoms in the brain: probing the mysteries of the human mind.* New York: William Morrow, 1998.

Ramachandran, V. S. *The tell-tale brain: a neuroscientist's quest for what makes us human.* New York: W. W. Norton, 2011.

Satel, Sally, and Scott O. Lilienfeld. *Brainwashed: the seductive appeal of mindless neuroscience.* New York: Basic Books, 2013.

Stien, Phyllis T., and Joshua C. Kendall. *Psychological trauma and the developing brain: neurologically based interventions for troubled children.* New York: Haworth Maltreatment and Trauma Press, 2004.

Introduction

Cheyne, James Allan, and Gordon Pennycook. "Sleep Paralysis Postepisode Distress." *Clinical Psychological Science* 1, no. 2 (2013): 135–48.

D'Agostino, Armando, and Ivan Limosani. "Hypnagogic Hallucinations and Sleep Paralysis." *Narcolepsy: a clinical guide.* New York: Springer, 2010.

Davies, Owen. "The Nightmare Experience, Sleep Paralysis, and Witchcraft Accusations." *Folklore* 114, no. 2 (2003): 181–203.

Santomauro, Julia, and Christopher C. French. "Terror in the Night." *The Psychologist* 22, no. 8 (2009): 672–75.

Chapter One: The Dueling Neurosurgeons

Baumgartner, Frederic J. *Henry II, king of France 1547–1559.* Durham: Duke University Press, 1988.

Faria, M. A. "The Death of Henry II of France." *Journal of Neurosurgery* 77, no. 6 (1992): 964–69.

Frieda, Leonie. *Catherine de Medici.* New York: Harper Perennial, 2006.

Goldstein, Lee E., et al. "Chronic Traumatic Encephalopathy in Blast-Exposed Military Veterans and a Blast Neurotrauma Mouse Model." *Science Translational Medicine* 4, no. 134 (2012): 134–60.

Keeton, Morris. "Andreas Vesalius: His times, his life, his work." *Bios* 7, no. 2 (1936): 97–109.

Martin, Graham. "The Death of Henry II of France: A sporting death and postmortem." *ANZ Journal of Surgery* 71, issue 5 (2001): 318–20.

Milburn, C. H. "An Address on Military Surgery of the Time of Ambroise Paré and That of the Present Time." *British Medical Journal* 1, no. 2112 (1901): 1532–35.

Miller, Greg. "Blast Injuries Linked to Neurodegeneration in Veterans." *Science* 336, no. 6083 (2012): 790–91.

O'Malley, Charles Donald. *Andreas Vesalius of Brussels, 1514–1564.* Berkeley, Calif.: University of California Press, 1964.

O'Malley, Charles Donald, and J. B. De C. M. Saunders, "The 'Relation' of Andreas Vesalius on the Death of Henry II of France." *Journal of the History of Medicine and Allied Sciences* 3, no. 1 (1948): 197–213.

Princess Michael of Kent. *The serpent and the moon: two rivals for the love of a Renaissance king.* New York: Simon & Schuster, 2004.

Rose, F. Clifford. "The History of Head Injuries: An overview." *Journal of the History of the Neurosciences* 6, no. 2 (1997): 154–80.
Simpson, D. "Paré as a Neurosurgeon." *The Australian and New Zealand Journal of Surgery* 67, no. 8 (1997): 540–46.
Strathern, Paul. *A brief history of medicine: from Hippocrates to gene therapy.* New York: Carroll & Graf, 2005.
Vesalius, Andreas, and J. B. de C. M. Saunders. *The illustrations from the works of Andreas Vesalius of Brussels.* Cleveland, Ohio: World, 1950.

Chapter Two: The Assassin's Soup

Ackerman, Kenneth D. *Dark horse: the surprise election and political murder of President James A. Garfield.* New York: Carroll & Graf, 2003.
De Carlos, Juan A., and José Borrell. "A Historical Reflection of the Contributions of Cajal and Golgi to the Foundations of Neuroscience." *Brain Research Reviews* 55, no. 1 (2007): 8–16.
Everett, Marshall. *Complete life of William McKinley and story of his assassination.* Cleveland, Ohio: N. G. Hamilton, 1901.
Finger, Stanley. *Minds behind the brain: a history of the pioneers and their discoveries.* Oxford: Oxford University Press, 2000.
Goldberg, Jeff. *Anatomy of a scientific discovery.* New York: Bantam Books, 1989.
Guiteau, Charles Julius, and C. J. Hayes. *A complete history of the trial of Guiteau, assassin of President Garfield.* Philadelphia: Hubbard Bros., 1882.
Haines, D. E. "Spitzka and Spitzka on the Brains of the Assassins of the Presidents." *Journal of the History of the Neurosciences* 4, no. 3/4 (1995): 236–66.
Johns, A. Wesley. *The man who shot McKinley.* South Brunswick, N.J.: A. S. Barnes, 1970.
Loewi, Otto. *An autobiographic sketch.* Chicago: University of Chicago, 1960.
Marcum, James A. "'Soup' vs. 'Sparks': Alexander Forbes and the synaptic transmission controversy." *Annals of Science* 63, no. 2 (2006): 139–56.
Menke, Richard. "Media in America, 1881: Garfield, Guiteau, Bell, Whitman." *Critical Inquiry* 31, no. 3 (2005): 638–64.
Miller, Scott. *The President and the assassin: McKinley, terror, and empire at the dawn of the American century.* New York: Random House, 2011.
Paulson, George. "Death of a President and his Assassin—Errors in Their Diagnosis and Autopsies." *Journal of the History of the Neurosciences* 15, no. 2 (2006): 77–91.
Peskin, Allan. "Charles Guiteau of Illinois, President Garfield's Assassin." *Journal of the Illinois State Historical Society* 70, no. 2 (1977): 130–39.
Rapport, Richard L. *Nerve endings: the discovery of the synapse.* New York: W. W. Norton, 2005.
Rauchway, Eric. *Murdering McKinley: the making of Theodore Roosevelt's America.* New York: Hill and Wang, 2003.

Sourkes, Theodore L. "The Discovery of Neurotransmitters, and Applications to Neurology." *Handbook of Clinical Neurology* 95, no. 1 (2009): 869–83.

University at Buffalo Libraries. "Pan-American Exposition of 1901." http://library.buffalo.edu/pan-am/ (accessed November 4, 2013).

Valenstein, Elliot S. *The war of the soups and the sparks.* New York: Columbia University Press, 2005.

Vowell, Sarah. *Assassination vacation.* New York: Simon & Schuster, 2005.

Chapter Three: Wiring and Rewiring

Brang, David, and V. S. Ramachandran. "Survival of the Synesthesia Gene: Why do people hear colors and taste words?" *PLoS Biology* 9, no. 11 (2011): 1–5.

Finkel, Michael. "The Blind Man Who Taught Himself to See." *Men's Journal*, March 2011. http://www.mensjournal.com/magazine/the-blind-man-who-taught-himself-to-see-20120504 (accessed November 4, 2013).

Fisher, Madeline. "Balancing Act." *On Wisconsin.* http://www.uwalumni.com/home/onwisconsin/archives/spring2007/balancingact.aspx (accessed November 4, 2013).

Hofmann, Albert. *LSD, my problem child.* New York: McGraw-Hill, 1980.

Holman, James. *A voyage round the world: including travels in Africa, Asia, Australasia, America, etc. etc. from MDCCCXXVII to MDCCCXXXII.* London: Smith, Elder, 1834.

Roberts, Jason. *A sense of the world.* New York: Harper Perennial, 2007.

Chapter Four: Facing Brain Damage

Alexander, Caroline. "Faces of War." *Smithsonian.* February 2007. http://www.smithsonianmag.com/history-archaeology/mask.html (accessed November 4, 2013).

Caramazza, Alfonso, and Jennifer R. Shelton. "Domain-Specific Knowledge Systems in the Brain." *Journal of Cognitive Neuroscience* 10, no. 1 (1998): 1–34.

Dubernard, Jean-Michel. "Outcomes 18 Months after the First Human Partial Face Transplant." *The New England Journal of Medicine* 357, no. 24 (2007): 2451–60.

Glickstein, Mitchell, and David Whitteridge. "Tatsuji Inouye and the Mapping of the Visual Fields on the Human Cerebral Cortex." *Trends in Neurosciences* 10, no. 9 (1987): 349–52.

Glickstein, Mitchell. "The Discovery of the Visual Cortex." *Scientific American.* September 1988: 118–27.

Hubel, David H. *Eye, brain, and vision.* New York: Scientific American Library, 1988.

———. "Evolution of Ideas on the Primary Visual Cortex, 1955–1978: A biased historical account." *Bioscience Reports* 2, no. 7 (1982): 435–69.

Khatchadourian, Raffi. "Transfiguration." *The New Yorker*. February 13 and 20, 2012: 66–87.

Moscovitch, Morris, Gordon Winocur, and Marlene Behrmann. "What is Special about Face Recognition?" *Journal of Cognitive Neuroscience* 9, no. 5 (1997): 555–604.

Nicolson, Juliet. *The great silence, 1918–1920: living in the shadow of the Great War*. London: John Murray, 2009.

Pinker, Steven. "So How Does the Mind Work?" *Mind & Language* 20, no. 1 (2005): 1–24.

Pomahac, Bohdan. "Three Patients with Full Facial Transplantations." *The New England Journal of Medicine* 366, no. 8 (2012): 715–22.

"Visual neuroscience: visual central pathways." *Visual neuroscience: visual central pathways*. http://camelot.mssm.edu/~ygyu/visualpathway.html (accessed August 15, 2012).

Chapter Five: The Brain's Motor

Anonymous. "The Case of George Dedlow." *The Atlantic Monthly*. July 1866.

Beatty, William K. "S. Weir Mitchell and the Ghosts." *Journal of the American Medical Association* 220, no. 1 (1972): 76–80.

Canale, D. J. "Civil War Medicine from the Perspective of S. Weir Mitchell's 'The Case of George Dedlow.'" *Journal of the History of the Neurosciences* 11, no. 1 (2002): 11–18.

———. "S. Weir Mitchell's Prose and Poetry on the American Civil War." *Journal of the History of the Neurosciences* 13, no. 1 (2004): 7–21.

Finger, Stanley, and Meredith P. Hustwit. "Five Early Accounts of Phantom Limb in Context: Paré, Descartes, Lemos, Bell, and Mitchell." *Neurosurgery* 52, no. 3 (2003): 675–86.

Freemon, Frank R. "The First Neurological Research Center: Turner's Lane hospital during the American Civil War." *Journal of the History of the Neurosciences* 2, no. 2 (1993): 135–42.

Goler, Robert I. "Loss and the Persistence of Memory: 'The Case of George Dedlow' and disabled Civil War veterans." *Literature and Medicine* 23, no. 1 (2004): 160–83.

Herschbach, Lisa. "'True Clinical Fictions': Medical and literary narratives from the Civil War hospital." *Culture, Medicine, and Psychiatry* 19, no. 2 (1995): 183–205.

Howey, Allan W. "The Rifle-Musket and the Minié Ball." *The Civil War Times*, October 1999. http://www.historynet.com/minie-ball (accessed November 4, 2013).

Lein, Glenna R. *The encyclopedia of Civil War medicine*. Armonk, N.Y.: M. E. Sharpe, 2008.

Mitchell, S. Weir. *Injuries of nerves and their consequences*. Philadelphia: J. B. Lippincott, 1872.

Ramachandran, Vilayanur S., and Diane Rogers-Ramachandran. "Synaesthesia in Phantom Limbs Induced with Mirrors." *Proceedings of the Royal Society of London, Biology* 263, no. 1369 (1996): 377–86.

———. "Phantom Limbs and Neural Plasticity." *Archives of Neurology* 57, no. 3 (2000): 317–20.

Ramachandran, Vilayanur S., and William Herstein. "The Perception of Phantom Limbs: The D. O. Hebb lecture." *Brain* 121, no. 9 (1998): 1603–20.

Chapter Six: The Laughing Disease

Anderson, Warwick. *The collectors of lost souls: turning kuru scientists into whitemen.* Baltimore: Johns Hopkins University Press, 2008.

Beasley, Anne. "Frontier Journeys: Fore experiences on the kuru patrols." *Oceania* 79, no. 1 (2009): 34–52.

"The End of Kuru: 50 years of research into an extraordinary disease." *Philosophical Transactions of the Royal Society B* 353, no. 1510 (2008): 3607–763.

Gajdusek, D. Carleton. *South Pacific expedition to the New Hebrides and to the Fore, Kukukuku, and Genatei peoples of New Guinea, January 26, 1967 to May 12, 1967.* Bethesda, Md.: Section of Child Growth and Development and Disease Patterns in Primitive Cultures, National Institute of Neurological Disease and Blindness, 1967.

The Genius and the Boys. DVD. Directed by Bosse Lindquist. Stockholm: SVT Documentary, 2009.

Georgopoulos, Apostolos P. "Movement, Balance, and Coordination—The Dana Guide." The Dana Foundation. http://www.dana.org/news/brainhealth/detail.aspx?id=10070 (accessed November 4, 2013).

Hainfellner, Johannes A., et al. "Pathology and Immunocytochemistry of a Kuru Brain." *Brain Pathology* 7, no. 1 (1997): 547–53.

Ledford, Heidi. "'Harmless' Prion Protein Linked to Alzheimer's Disease." Nature.com. http://www.nature.com/news/2009/090225/full/news.2009.121.html (accessed November 4, 2013).

Lindenbaum, Shirley. "Kuru, Prions, and Human Affairs: Thinking about epidemics." *Annual Review of Anthropology* 30, no. 1 (2001): 363–85.

Miller, Greg. "Could They All Be Prion Diseases?" *Science* 326, no. 5958 (2009): 1337–39.

Nelson, Hank. "Kuru: The Pursuit of the Prize and the Cure." *The Journal of Pacific History* 31, no. 2 (1996): 178–201.

Norrby, Erling. *Nobel prizes and life sciences.* Singapore: World Scientific, 2010.

Spark, Geridwen. "Carleton's Kids." *The Journal of Pacific History* 44, no. 1, (2009): 1–19.

Stern, Nicholas C. "Agents Investigated Nobel Prize Winner Daniel Gajdusek as Far Back as 1950s." *Frederick News-Post,* October 25, 2009. http://www.fredericknewspost.com/archive/article_9c620533-8d25-5ed5-8a53-96ac409697f5.html?mode=story (accessed November 4, 2013).

Chapter Seven: Sex and Punishment

Batts, Shelley. "Brain Lesions and Their Implications in Criminal Responsibility." *Behavioral Science and the Law* 27, no. 2 (2009): 261–72.

Bliss, Michael. *Harvey Cushing: a life in surgery.* New York: Oxford University Press, 2005.

Byrne, John H. *Learning and memory.* New York: Macmillan Reference USA, 2003.

Cushing, Harvey. "Partial Hypophysectomy for Acromegaly." *Annals of Surgery* 50, no. 6 (1909): 1002–17.

———. *The pituitary body and its disorders.* Philadelphia and London: J. B. Lippincott, 1912.

Damasio, Antonio R., Daniel Tranel, and Helen Damasio. "Individuals with Sociopathic Behavior Caused by Frontal Damage Fail to Respond Autonomically to Social Stimuli." *Behavioural Brain Research* 41, no. 2 (1990): 81–94.

Damasio, Antonio R. *Descartes' error: emotion, reason, and the human brain.* New York: Putnam, 1994.

Denzel, Justin F. *Genius with a scalpel.* New York: Messner, 1971.

Devinsky, Julie, Oliver Sacks, Orrin Devinsky. "Kluver-Bucy Syndrome, Hypersexuality, and the Law." *Neurocase* 16, no. 2 (2010): 140–45.

Devinsky, Orrin. *Neurology of cognitive and behavioral disorders.* Oxford: Oxford University Press, 2004.

Eslinger, Paul J., and Antonio R. Damasio. "Severe Disturbance of Higher Cognition after Bilateral Frontal Lobe Ablation: Patient EVR." *Neurology* 35, no. 12 (1985): 1731–41.

Feinstein, Justin S., et al. "The Human Amygdala and the Induction and Experience of Fear." *Current Biology* 21, no. 1 (2010): 34–38.

Fulton, John F. *Harvey Cushing: a biography.* New York: Arno Press, 1980.

Greenwood, Richard, et al. "Behaviour Disturbances During Recovery from Herpes Simplex Encephalitis." *Journal of Neurology, Neurosurgery, and Psychiatry* 46, no. 9 (1983): 809–17.

Kalat, James W., and Michelle N. Shiota. *Emotion.* Belmont, Calif.: Thomson Wadsworth, 2007.

Lehrer, Steven. *Explorers of the body.* Garden City, N.Y.: Doubleday, 1979.

Morte, Paul D. "Neurologic Aspects of Human Anomalies." *Western Journal of Medicine* 139, no. 2 (1983): 250–56.

Papez, James Wenceslaus. "A Proposed Mechanism of Emotion." *Archives of Neurology and Psychiatry* 38, no. 4 (1937): 725–43.

Sapolsky, Robert M. "The Frontal Cortex and the Criminal Justice System." *Philosophical Transactions of the Royal Society B* 359, no. 1451 (2004): 1787–96.

Sharpe, William. *Brain surgeon: the autobiography of William Sharpe.* London: Gollancz, 1953.

Chapter Eight: The Sacred Disease

Cassano, D. "Neurology and the Soul: From the origins until 1500." *Journal of the History of the Neurosciences* 5, no. 2 (1996): 152–61.

Costandi, Mo. "Diagnosing Dostoyevsky's Epilepsy." Science Blogs. http://neurophilosophy.wordpress.com/2007/04/16/diagnosing-dostoyevskys-epilepsy/ (accessed November 4, 2013).

———. "Wilder Penfield, Neural Cartographer." Science Blogs. http://scienceblogs.com/neurophilosophy/2008/08/27/wilder-penfield-neural-cartographer/ (accessed November 4, 2013).

Eccles, John. "Wilder Graves Penfield." In *Memoirs of Fellows of the Royal Society.* London: The Royal Society, 1978: 473–513.

Finger, Stanley. "The Birth of Localization Theory." *Handbook of Clinical Neurology* 95, no. 1 (2010): 117–28.

Foote-Smith, Elizabeth, and Timothy J. Smith. "Historical Note: Emanuel Swedenborg." *Epilepsia* 37, no. 2 (1996): 211–18.

Harris, Lauren Julius, and Jason B. Almerigi. "Probing the Human Brain with Stimulating Electrodes: The story of Roberts Bartholow's (1874) experiment on Mary Rafferty." *Brain and Cognition* 70, no. 1 (2009): 92–115.

Hebb, Donald O., and Wilder Penfield. "Human Behavior after Extensive Bilateral Removal from the Frontal Lobes." *Archives of Neurology and Psychiatry* 44, no. 2 (1940): 421–38.

Jones, Simon R. "Talking Back to the Spirits: The voices and visions of Emanuel Swedenborg." *History of the Human Sciences* 21, no. 1 (2008): 1–31.

Lewis, Jefferson. *Something hidden: a biography of Wilder Penfield.* Toronto, Ont.: Doubleday Canada, 1981.

Morgan, James P. "The First Reported Case of Electrical Stimulation of the Human Brain." *Journal of the History of Medicine and Allied Sciences* 37, no. 1 (1982): 51–64.

Newberg, Andrew B., Eugene G. Aquili, and Vince Rause. *Why God won't go away: brain science and the biology of belief.* New York: Ballantine Books, 2001.

Penfield, Wilder. "The Frontal Lobe in Man: A clinical study of maximal removals." *Brain* 58, no. 1 (1935): 115–33.

———. "The Interpretive Cortex." *Science* 129, no. 3365 (1959): 1719–25.

———. *No man alone: a neurosurgeon's life.* Boston: Little, Brown, 1977.

———. "Some Mechanisms of Consciousness Discovered During Electrical Stimulation of the Brain." *Proceedings of the National Academy of Science* 44, no. 2 (1958): 51–66.

———. "The Twenty-Ninth Maudsley Lecture: The role of the temporal cortex in certain psychical phenomenon." *Journal of Mental Science* 101, no. 424 (1955): 451–65.

Taylor, Charlotte S. R., and Charles G. Gross. "Twitches Versus Movements: A story of motor cortex." *Neuroscientist* 9, no. 5 (2003): 332–42.

Zago, Stefano, et al. "Bartholow, Sciamanna, Alberti: Pioneers in the electrical stimulation of the exposed human cerebral cortex." *Neuroscientist* 14, no. 5 (2008): 521–28.

Chapter Nine: "Sleights of Mind"

Biran, Iftah, et al. "The Alien Hand Syndrome: What makes the alien hand alien?" *Cognitive Neuropsychology* 23, no. 4 (2006), 563–82.

Breen, Nora, et al. "Towards an Understanding of Delusions of Misidentification: Four case studies." *Mind and Language* 15, no. 1 (2000): 74–110.

Cooper, John Milton. *Woodrow Wilson: a biography.* New York: Alfred A. Knopf, 2009.

Custers, R., and H. Aarts. "The Unconscious Will: How the pursuit of goals operates outside of conscious awareness." *Science* 329, no. 5987 (2010): 47–50.

Draaisma, Douwe. "Echoes, Doubles, and Delusions: Capgras syndrome in science and literature." *Style* 43, no. 3 (2009): 429–41.

Ellis, Hadyn D., and Michael B. Lewis. "Capgras Delusion: A window on face recognition." *Trends in Cognitive Science* 5, no. 4, (2001): 149–56.

Ellis, Hadyn D., et al. "Reduced Autonomic Responses for Faces in Capgras Delusion." *Proceedings of the Royal Society of London B* 264, no. 1384 (1997): 1085–92.

Fisher, C. M. "Alien Hand Phenomena: A review of the literature with the addition of six personal cases." *Canadian Journal of Neurological Sciences* 27, no. 3 (2000): 192–203.

Hirstein, William, and Vilayanur S. Ramachandran. "Capgras Syndrome: A novel probe for understanding the neural representation of the identity and familiarity of persons." *Proceedings of the Royal Society of London B* 264, no. 1380 (1997): 437–44.

Klemm, W. R. "Free Will Debates: Simple experiments are not so simple." *Advances in Cognitive Psychology* 6, no. 1 (2010): 47–65.

Libet, Benjamin, et al. "Time of Conscious Intention to Act in Relation to Onset of Cerebral Activity (Readiness-Potential)." *Brain* 106, no. 3 (1983): 623–42.

McKay, Ryan, Robyn Langdon, and Max Coltheart. " 'Sleights of Mind': Delusions, defenses, and self-deception." *Cognitive Neuropsychiatry* 10, no. 4 (2005): 305–26.

Morris, Errol. "The Anosognosic's Dilemma: Something's wrong but you'll never know what it is." *New York Times.* http://opinionator.blogs.nytimes.com/2010/06/20/the-anosognosics-dilemma-1/ (accessed November 4, 2013).

Prigatano, George P., and Daniel L. Schacter. *Awareness of deficit after brain injury: clinical and theoretical issues.* New York: Oxford University Press, 1991.

Ramachandran, Vilayanur S. "What Neurological Syndromes Can Tell Us about Human Nature." *Cold Spring Harbor Symposium on Quantitative Biology* 61, no. 1 (1996): 115–34.

Roskies, Adina. "Neuroscientific Challenges to Free Will and Responsibility." *Trends in Cognitive Science* 10, no. 9 (2006): 419–23.
Scepkowski, Lisa A., and Alice Cronin-Golomb. "The Alien Hand: Cases, categorizations, and anatomical correlates." *Behavioral and Cognitive Neuroscience Reviews* 2, no. 4 (2003): 261–77.
Todd, J. "The Syndrome of Alice in Wonderland." *Canadian Medical Association Journal* 73, no. 9 (1955): 701–4.
Weinstein, Edwin A. "Woodrow Wilson's Neurological Illness." *Journal of American History* 57, no. 2 (1970): 324–51.

Chapter Ten: Honest Lying

Berrios, German E. "Confabulations: A conceptual history." *Journal of the History of the Neurosciences* 7, no. 3 (1998): 225–41.
Bruyn, G. W., and Charles M. Poser. *The history of tropical neurology: nutritional disorders.* Canton, MA: Science History Publications/USA, 2003.
Buckner, Randy L., and Mark E. Wheeler. "The Cognitive Neuroscience of Remembering." *Nature Reviews: Neuroscience* 2, no. 9 (2001): 624–34.
Corkin, Suzanne. "Lasting Consequences of Bilateral Medial Temporal Lobectomy." *Seminars in Neurology* 4, no. 2 (1984): 249–59.
———. *Permanent present tense: the unforgettable life of the amnesic patient, H.M.* New York: Basic Books, 2013.
———. "What's New with Amnesic Patient H.M.?" *Nature Reviews: Neuroscience* 3, no. 2 (2002): 153–60.
Dalla Barba, Gianfranco. "Consciousness and Confabulation: Remembering 'another' past." In *Broken memories: case studies in memory impairment.* Oxford: Blackwell, 1995: 101–23.
De Wardener, Hugh Edward, and Bernard Lennox. "Cerebral Beriberi (Wernicke's Encephalopathy): Review of 52 cases in a Singapore prisoner-of-war hospital." *The Lancet* 249, no. 6436 (1947): 11–17.
Dittrich, Luke. "The Brain That Changed Everything." *Esquire.* http://www.esquire.com/features/henry-molaison-brain-1110 (accessed November 4, 2013).
Gabrieli, J. D. E. "Cognitive Neuroscience of Human Memory." *Annual Review of Psychology* 49, no. 1 (1998): 87–115.
Hirstein, William. "What Is Confabulation?" In *Brain fiction: self-deception and the riddle of confabulation.* Cambridge, Mass.: MIT Press, 2005: 1–23.
Klein, Stanley B., Keith Rozendal, and Leda Cosmides. "A Social-Cognitive Neuroscience Analysis of the Self." *Social Cognition* 20, no. 2 (2002): 105–35.
Kopelman, Michael D. "Disorders of Memory." *Brain* 125, no. 10 (2002), 2152–90.
Luria, A. R. *The mind of a mnemonist.* New York: Penguin, 1975.
MacLeod, Sandy. "Psychiatry on the Burma-Thai Railway (1942–1943): Dr. Rowley Richards and colleagues." *Australasian Psychiatry* 18, no. 6 (2010): 491–95.

Martin, Peter R., Charles K. Singleton, and Susanne Hiller-Sturmhöfel. "The Role of Thiamine Deficiency in Alcoholic Brain Disease." *Alcohol Research and Health* 27, no. 2 (2003): 134–42.

McGaugh, James L. "Memory: A century of consolidation." *Science* 287, no. 5451 (2000): 248–51.

Postle, Bradley R. "The Hippocampus, Memory, and Consciousness." In *The neurology of consciousness: cognitive neuroscience and neuropathology*. Amsterdam: Elsevier Academic Press, 2009: 326–38.

Rosenbaum, R. Shayna, et al. "The Case of K.C.: Contributions of a memory-impaired person to memory theory." *Neuropsychologia* 43, no. 7 (2005): 989–1021.

Scoville, William B., and Brenda Milner. "Loss of Recent Memory after Bilateral Hippocampal Lesions." *Journal of Neurology, Neurosurgery, and Psychiatry* 20, no. 1 (1957): 11–21.

Van Damme, Ilse, and Géry d'Ydewalle. "Confabulation Versus Experimentally Induced False Memories in Korsakoff's Syndrome." *Journal of Neuropsychology* 4, no. 2 (2010): 211–30.

Wilson, Barbara A., Michael Kopelman, and Narinder Kapur. "Prominent and Persistent Loss of Past Awareness in Amnesia." *Neuropsychological Rehabilitation* 18, no. 5–6 (2008): 527–40.

Xia, Chenjie. "Understanding the Human Brain: A lifetime of dedicated pursuit." *McGill Journal of Medicine* 9, no. 2 (2006): 165–72.

Zannino, Gian Daniele, et al. "Do Confabulators Really Try to Remember When They Confabulate? A Case Report." *Cognitive Neuropsychology* 25, no. 6 (2008): 831–52.

Chapter Eleven: Left, Right, and Center

Berlucchi, Giovanni. "Revisiting the 1981 Nobel Prize to Roger Sperry, David Hubel, and Torsten Wiesel on the Occasion of the Centennial of the Prize to Golgi and Cajal." *Journal of the History of the Neurosciences* 15, no. 4 (2006): 369–75.

Borod, Joan C., Cornelia Santschi Haywood, and Elissa Koff. "Neuropsychological Aspects of Facial Asymmetry During Emotional Expression: A review of the normal adult literature." *Neuropsychological Review* 7, no. 1 (1997): 41–60.

Broca, Paul. "On the Site of the Faculty of Articulated Speech (1865)." *Neuropsychology Review* 21, no. 3 (2011): 230–35.

Broca, Paul, and Christopher D. Green (trans.). "Remarks on the Seat of the Faculty of Articulated Language, Following an Observation of Aphemia (Loss of Speech) (1861)." *Classics in the History of Psychology*. http://psychclassics.yorku.ca/Broca/aphemie-e.htm (accessed November 4, 2013).

Christiansen, Morten H., and Nick Chater. "Language as Shaped by the Brain." *Behavioral and Brain Sciences* 31, no. 5 (2008): 489–508.

Critchley, Macdonald. "The Broca-Dax Controversy." In *The divine banquet of the brain and other essays.* New York: Raven Press, 1979: 72–82.

Engel, Howard. *The man who forgot how to read.* New York: Thomas Dunne Books/St. Martin's Press, 2008.

Gazzaniga, Michael S. "Cerebral Specialization and Interhemispheric Communication." *Brain* 123, no. 7 (2000): 1293–326.

———. "Forty-Five Years of Split-Brain Research and Still Going Strong." *Nature Reviews: Neuroscience* 6, no. 8 (2005): 653–59.

———. "The Split Brain in Man." *Scientific American* 217, no. 2 (1967): 24–29.

Gazzaniga, Michael S., Joseph E. Bogan, and Roger W. Sperry. "Observations on Visual Perception After Disconnexion of the Cerebral Hemispheres in Man." *Brain* 88, no. 2 (1965): 221–36.

———. "Some Functional Effects of Sectioning the Cerebral Commissures in Man." *Proceedings of the National Academy of Sciences* 48, no. 10 (1962): 1765–69.

Gazzaniga, Michael S., et al. "Neurologic Perspectives on Right Hemisphere Language Following Surgical Section of the Corpus Callosum." *Seminars in Neurology* 4, no. 2 (1984): 126–35.

Henderson, Victor W. "Alexia and Agraphia." *Handbook of Clinical Neurology* 95, no. 1 (2009): 583–601.

MacNeilage, Peter F., Lesley J. Rogers, and Giorgio Vallortigara. "Origins of the Left and Right Brain." *Scientific American.* June 24, 2009.

Schiller, Francis. *Paul Broca, founder of French anthropology, explorer of the brain.* Berkeley: University of California Press, 1979.

Skinner, Martin, and Brian Mullen. "Facial Asymmetry in Emotional Expressions: A meta-analysis of research." *British Journal of Social Psychology* 30, no. 2 (1991): 113–24.

Sperry, Roger. "Some Effects of Disconnecting the Cerebral Hemispheres." *Science* 217, no. 4566 (1982): 1223–26.

Wolman, David. "A Tale of Two Halves." *Nature* 483, no. 7389 (2012): 260–63.

Chapter Twelve: The Man, the Myth, the Legend

Alvarez, Julie A., and Eugene Emory. "Executive Function and the Frontal Lobes." *Neuropsychological Review* 16, no. 1 (2006): 17–42.

Devinsky, Orrin. "Executive Function and the Frontal Lobes." In *Neurology of cognitive and behavioral disorders.* Oxford: Oxford University Press, 2004: 302–29.

Dominus, Susan. "Could Conjoined Twins Share a Mind?" *New York Times Magazine,* May 29, 2011: MM28–35.

Gordon, D. S. "Penetrating Head Injuries." *Ulster Medical Journal* 57, no. 1 (1988): 1–10.

Harmon, Katherine. "The Chances of Recovering from Brain Trauma: Past cases show why millimeters matter." *Scientific American.* http://www.scientificamerican.com/article.cfm?id=recovering-from-brain-trauma (accessed November 4, 2013).

Kotowicz, Zbigniew. "The Strange Case of Phineas Gage." *History of the Human Sciences* 20, no. 1 (2007): 115–31.

Macmillan, Malcolm. "Phineas Gage—Unraveling the Myth." *Psychologist* 21, no. 9 (2008): 828–31.

Macmillan, Malcolm, and Matthew L. Lena. "Rehabilitating Phineas Gage." *Neuropsychological Rehabilitation* 20, no. 5 (2010): 641–58.

Sacks, Oliver. "The Abyss." *The New Yorker*, September 24, 2007: 100–108.

Stone, James L. "Transcranial Brain Injuries Caused by Metal Rods or Pipes Over the Past 150 Years." *Journal of the History of the Neurosciences* 8, no. 3 (1999): 227–34.

Wearing, Deborah. *Forever today: a memoir of love and amnesia.* London: Doubleday, 2005.

Wilson, Barbara A., and Deborah Wearing. "Prisoner of Consciousness: A state of just awakening following herpes simplex encephalitis." In *Broken memories: case studies in memory impairment.* Oxford: Blackwell, 1995: 14–30.

Wilson, Barbara A., Alan D. Braddeley, and Narinder Kapur. "Dense Amnesia in a Professional Musician Following Herpes Simplex Virus Encephalitis." *Journal of Clinical and Experimental Neuropsychology* 17, no. 5 (1995): 668–81.

Wilson, Barbara A., Michael Kopelman, and Narinder Kapur. "Prominent and Persistent Loss of Past Awareness in Amnesia." *Neuropsychological Rehabilitation* 18, no. 5–6 (2008): 527–40.

Wilgus, Jack, and Beverly Wilgus. "Face to Face with Phineas Gage." *Journal of the History of the Neurosciences* 18, no. 3 (2009): 340–45.

Index